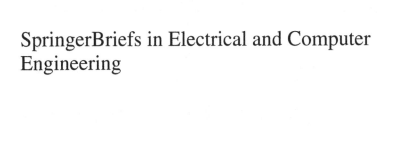

SpringerBriefs in Electrical and Computer Engineering

More information about this series at http://www.springer.com/series/10059

Khadige Abboud • Weihua Zhuang

Mobility Modeling for Vehicular Communication Networks

Springer

Khadige Abboud
University of Waterloo
Waterloo, ON, Canada

Weihua Zhuang
Department of Electrical
    and Computer Enginnering
University of Waterloo
Waterloo, ON, Canada

ISSN 2191-8112   ISSN 2191-8120   (electronic)
SpringerBriefs in Electrical and Computer Engineering
ISBN 978-3-319-25505-7   ISBN 978-3-319-25507-1   (eBook)
DOI 10.1007/978-3-319-25507-1

Library of Congress Control Number: 2015954177

Springer Cham Heidelberg New York Dordrecht London

Printed on acid-free paper

Springer International Publishing AG Switzerland is part of Springer Science+Business Media (www.
springer.com)

To my parents—K.A.

To Alan and Alvin—W.Z.

Preface

In recent years, efforts have been made to deploy communication capabilities in vehicles and the transport infrastructure, leading to a potential of vehicular ad hoc networks (VANETs). Enabling vehicular communications is expected to revolutionize the transport infrastructure and support not only public-safety applications, but also a wide range of infotainment applications. Urban roads and highways are highly susceptible to high relative speeds and traffic density variations from time-to-time and from point-to-point on the same roads. High node mobility in VANETs can cause frequent network topology changes and fragmentations. The spatiotemporal variations in VANET topology directly (or indirectly) affect the performance of network protocols through their impact on the switching of the links between connection and disconnection. Characterizing the spatiotemporal variations in VANET topology requires a mathematical model that describes vehicle mobility and its temporal variations, that is, a microscopic mobility model. Despite its high speed and randomness, the vehicle movement is restricted by road topology, speed limits, traffic rules, and movement of nearby vehicles. Therefore, vehicle movement follows certain patterns. This brief presents a stochastic microscopic mobility model that describes the temporal changes of intervehicle distances, accounting for the dependency between these changes. The model is consistent with simulated and empirical vehicle traffic patterns. Using stochastic lumpability method, the proposed mobility model is mapped into an aggregated mobility model that describes the mobility of a group of vehicles. In addition, the proposed mobility model is utilized to analyze the spatiotemporal VANET topology. Two characteristics are proposed to describe the impact of vehicle mobility on VANET topology: the time period between successive changes in communication link state (connection and disconnection) and the time period between successive changes in node's one-hop neighborhood. Using the proposed lumped group mobility model, the two VANET topology characteristics are probabilistically described for different vehicular traffic flow conditions. Furthermore, the limiting behavior of a system of two-hop vehicles is investigated in terms of the overlap-state of their coverage ranges and the number of common vehicle neighbors between them. The steady-state number of common vehicle neighbors between two-hop vehicles is approximately derived. The proposed mobility model is a promising candidate model to be utilized for traceable mathematical analysis in VANETs. The spatiotemporal VANET topology analysis provides a useful tool for the development of mobility-aware vehicular network protocols.

Waterloo, ON, Canada   Khadige Abboud
                                                                         Weihua Zhuang

Acknowledgments

We would like to thank all our colleagues in the Broadband Communications Research (BBCR) Lab at the University of Waterloo for the fruitful discussions, research collaboration, and continuous exchange of knowledge.

Contents

Acronyms

1D	1-dimensional
cdf	Cumulative distribution function
CCH	Control channel
DSRC	Dedicated Short Range Communications
ETSI	European Telecommunications Standards Institute
FCC	Federal Communications Commission of the U.S.
i.i.d.	Independent and identically distributed
ITS	Intelligent Transportation Systems
LR	Linear regression
MAC	Medium Access Control
NGSIM	Next generation simulation
NHTSA	National Highway Traffic Safety Administration of the U.S.
pdf	Probability density function
pmf	Probability mass function
r.v.	Random variable
SCHs	Service channels
s.p.	Stochastic process
s.t.	Such that
U.S.	United States (of America)
VN	Vehicle neighbor

Nomenclature

$\prod$	Product notation
$\sum$	Sum notation
$\{\}$	Set notation
$()$	Sequence notation
α	Location parameter for the distribution of the inter-vehicle distance (the minimum distance headway) in meters
β	State dependency parameter of the probability transition matrix
$\Gamma(x)$	Gamma function
$\gamma(z, x)$	The lower incomplete gamma function
$\Gamma(z, x)$	The upper incomplete gamma function
δ_j	Absorbing probability of lumped state $\Omega_j \in \Omega_R$
δ_{E_e}	Absorbing probability of lumped state $\Omega_e \in \Omega_E$
$\delta_{E'_e}$	Absorbing probability δ_{E_e} weighted by the stationary distribution π_i
δ_{O_o}	Absorbing probability of lumped state $\Omega_o \in \Omega_O$
ζ_i	The length of the ith overlapping period (r.v.)
η_i	The length of the ith non-overlapping period (r.v.)
$\theta(m)$	A stochastic process with state space $\{-1, 1\}$ representing the range-overlap state (s.p.)
Θ_E	A set of states of the edge lumped Markov chain for system $\mathbb{X}_{VN}$ corresponding to initial state of the system
λ	Scale parameter for the mesoscopic probability distribution of the distance headway in near-capacity traffic flow conditions
$\lambda_u^{(j)}$	The uth non-unit eigenvalue of $M^{(j)}$
λ_u	The uth non-unit eigenvalue of M
μ	Mean distance headway in meters
ξ_{k-1}	The service time of customer $k - 1$ (r.v.)
π_i	Stationary distribution of the ith state in the 1D-Markov chain
σ	Standard deviation of the distance headway in meters
τ	The duration of the time step in the Markov chain model in seconds
ϕ_i	Absorbing probability of lumped state $\Omega_i \in \Omega_R$

ψ_i	Absorbing probability of lumped state $\Omega_i \in \Omega_I$
Ω_E	A set of lumped states in the Edge lumped Markov chain of system $\mathbb{X}_{VN}$ corresponding to a node entering the coverage range of the reference vehicle
Ω_i	The ith lumped state in the lumped Markov chain.
Ω_I	A set of lumped states in the lumped Markov chain such that the system $\mathbb{X}_N$ is not in the event set
Ω_L	A set of lumped states in the lumped Markov chain indicating the occurrence of a certain event of interest to system $\mathbb{X}_N$
Ω_O	A set of lumped states in the Edge lumped Markov chain of system $\mathbb{X}_{VN}$ corresponding to a node leaving the coverage range of the reference vehicle
ϱ	The permutation operator that maps two sequences under a specific index order.
$\mho_i$	Stationary distribution of lumped state Ω_i
A_{k-1}	The interarrival time between customers $k - 1$ and k (r.v.)
B_k	The buffer content at the beginning of the kth cycle (r.v.)
$c_{T_{li}}$	The coefficient of variation of T_{li}
$c_{T_{lo}}$	The coefficient of variation of T_{lo}
$C(l)$	The event that there exists at least one node within distance l from a reference node
$C^c(l)$	The event that there are no nodes within distance l from a reference node (the complement of event $C(l)$)
D	Vehicle density in vehicle per kilometer per lane (veh/km/lane)
$\mathrm{erf}(\cdot)$	The error function
e_{i_l}	The event that a vehicle enters the coverage range from the left side of the reference vehicle
$e_{i_l 2}$	The event that a vehicle enters the leading vehicle's coverage range from its left side
e_{i_r}	The event that a vehicle enters the coverage range from the right side of the reference vehicle
$e_{i_r 1}$	The event that a vehicle enters the following vehicle's coverage range from its right side
e_{o_l}	The event that a vehicle leaves the coverage range from the left side of the reference vehicle
e_{o_r}	The event that a vehicle leaves the coverage range from the right side of the reference vehicle
$F_Y(y)$	The cumulative distribution function of random variable Y
$f_Y(y)$	The probability density function of random variable Y
H	The length of the communication hop from a reference node in meters (r.v.)
I_0	The super state of system $\mathbb{X}_{N_H}$ at the 0th time step
L_s	The length of the range covered by each state in the 1D-Markov chain measured in meters

M	Probability transition matrix of the 1D-Markov chain
M'	The probability transition matrix of the absorbing 1D-Markov chain
M_N	The probability transition matrix of the lumped Markov chain representing the system $\mathbb{X}_N$
$\tilde{M}_{N_H}$	The probability transition matrix of the absorbing lumped Markov chain representing the system $\mathbb{X}_{N_H}$, when lumped states in Ω_R are merged into a single absorbing state
M''_{N_H}	The probability transition matrix of the absorbing lumped Markov chain representing the system $\mathbb{X}_{N_H}$, when lumped states in Ω_R are made absorbing without merging them into a single state
$\ddot{M}_N$	The transition matrix of the N-dimensional Markov chain that represents the system $\mathbb{X}_N$
$M_{N_{VN}}$	The probability transition matrix of the edge lumped Markov chain describing the system $\mathbb{X}_{VN}$
$\tilde{M}_{N_{VN}}$	The probability transition matrix of the absorbing edge lumped Markov chain, when lumped states in Ω_O and Ω_E are merged into one single absorbing state
$\tilde{M}_{N_{H2}}$	The probability transition matrix of the absorbing lumped Markov chain representing the system $\mathbb{X}_{N_{H2}}$, when lumped states in Ω_{2R} are merged into a single absorbing state
$M'_{N_{VN}}$	The probability transition matrix of the absorbing edge lumped Markov chain, when lumped states in Ω_O and Ω_E are made absorbing without merging them into a single state
N	An arbitrary number of distance headways between two reference vehicles
N_H	The number of distance headways between a reference node and its hop edge node at the 0th time step
$N_i(\zeta_k)$	The number of nodes entering the buffer during the kth range-overlapping period (r.v.)
$N_i(\Delta t)$	The number of nodes entering the overlapping region during an arbitrary time period, Δt (r.v.)
$N_{D,i}$	The number of distinct states in the lumped state Ω_i
$n_{j,j'}$	The number of transitions of a distance headway from state j to state j' within a time step of length τ in the NGSIM/VISSIM trajectory data
n_j	The number of time steps at which the distance headway is in state j in the NGSIM/VISSIM trajectory data
N_L	The state space size of the lumped Markov chain
$\tilde{N}_L$	The state space size of the absorbing lumped Markov chain
$N_{\max}$	Number of states in the 1D-Markov chain
$N_o(\theta_k)$	The number of nodes leaving the buffer during the kth range-non-overlapping period (r.v.)
$N_o(\Delta t)$	The number of nodes leaving the uncovered region during an arbitrary time period, Δt (r.v.)
N_R	The integer number of the states that cover distance headways within R in the distance headway's 1D-Markov chain

N_{th}	A threshold for the sum of indices of states of system $\mathbb{X}_N$ that is predefined for a certain event of interest
N_{VN}	The number of VNs on one side of a reference vehicle (r.v.)
O_{ij}	The index order that is applied on sequence S_i to get sequence S_j.
$O(.)$	The big O notation as a measure of computational complexity
$O_E(\Omega_i)$	A function that maps a lumped state from edge lumped markov chain to the corresponding one in the fully markov chain
p	Density dependent parameter for the transition probability from state j to $j+1$ in the 1D-Markov chain
P_{C0}	The limiting probability that there is zero common VNs between two reference vehicles
$p_{j,j'}$	The transition probability from state j to j' in the NGSIM/VISSIM trajectory data
p_j	The transition probability from state j to state $j+1$ in the 1D-Markov chain within one time step
P_{nov}	The limiting range-overlapping probability
P_{ov}	The limiting range-non-overlapping probability
P_{U0}	The limiting probability that there is zero uncovered nodes between two reference vehicles
q	Density dependent parameter for the transition probability from state j to $j-1$ in the 1D-Markov chain
q_j	The transition probability from state j to state $j-1$ in the 1D-Markov chain within one time step
R	Transmission range in meters
r_j	Return probability of state j back to itself in the 1D-Markov chain within one time step
s_i	i^{th} state in the the 1D-Markov chain
S_i	i^{th} super state in the the N-dimensional Markov chain
$|\mathbb{S}_N|$	The state space size of the Markov chain representing the system $\mathbb{X}_N$
$T(e)$	The first occurrence time of event e (r.v.)
T_i	The first arrival time of nodes to the overlapping region (r.v.)
T_{Ii}	The interarrival time of nodes to the overlapping region (r.v.)
T_{Io}	Node interdeparture time from the uncovered region that causes the number of uncovered nodes to decrease (r.v.)
$T^i_{j,j'}$	The first passage time of the distance headway X_i to state j' given that the distance headway is in state j at the 0th time step
T_L	The time interval that the system spends in the initial set Ω_I
T_{nov}	Range-non-overlapping time period (r.v.)
T_{ov}	Range-overlapping time period (r.v.)
$T_{R0}(\Omega_k)$	The time period for the first link breakage between a vehicle and its hop edge node, given that the distance headways between them are initially in states $I_0 \in \Omega_k$
T_{R0}	The time period for the first link breakage between a vehicle and its hop edge node

T_R	The communication link lifetime
T_{VN}	The time interval between two consecutive changes in vehicle-neighborhood (r.v.)
T_{VN0}	The time for the first change in vehicle-neighborhood to occur after initial time (0th time step) (r.v.)
$T_{VN0}(\Omega_k)$	Time for the first change in vehicle-neighborhood to occur after initial time (0th time step, given that system $\mathbb{X}_{VN}$ is initially in state Ω_k (r.v.)
T_{VN0_l}	First occurrence time of the first change in vehicle-neighborhood to occur (after initial time (0th time step) due to a vehicle leaving and entering the coverage range from the left side of the reference vehicle (r.v.)
T_{VN0_r}	First occurrence time of the first change in vehicle-neighborhood to occur (after initial time (0th time step) due to a vehicle leaving and entering the coverage range from the right side of the reference vehicle (r.v.)
$\bar{v}$	The maximum relative speed between vehicles in meters per second
X_i	The distance headway between node i and node $i + 1$ in meters (s.p.)
$X_i(m)$	The distance headway between node i and node $i + 1$ in meters at the mth (r.v.)
x_j	The quantized distance headway length of the jth state in the 1D-Markov chain measured in meters
$\mathbb{X}_N$	Sequence of (N) distance headways between two reference vehicles (s.p.)
$\mathbb{X}_{N_H}$	Sequence of distance headways between a reference vehicle and its hop edge node (s.p.)
$\mathbb{X}_{VN}$	Sequence of distance headways of the reference vehicle and the N_{VN} nodes in the coverage range on one side of the reference vehicle
$\mathbb{X}_{VN_E}$	Sequence of distance headways of the reference vehicle and its VNs when the first vehicle-neighborhood change occurs due to a node entering the reference vehicle's coverage range (s.p)
$\mathbb{X}_{VN_O}$	Sequence of distance headways of the reference vehicle and its VNs when the first vehicle-neighborhood change occurs due to a node leaving the reference vehicle's coverage range (s.p)
z	Shape parameter for the mesoscopic probability distribution of the distance headway in near-capacity traffic flow conditions

Chapter 1
Introduction

1.1 Vehicular ad hoc Networks (VANETs)

Newly manufactured vehicles are no longer the simple mechanical devices that
we once knew. Each vehicle is a smart body of various sensors that can measure
different attributes. Recently, efforts have been made to deploy communication
capabilities in vehicles and the transport infrastructure, leading to a potential of
vehicular ad hoc networks (VANETs) [1–3]. In 1999, the United States Federal
Communications Commission (FCC) allocated 75 MHz of radio spectrum in the
5.9 GHz band to be used for Dedicated Short Range Communication (DSRC) by
intelligent transportation systems (ITS). The DSRC spectrum has seven 10MHz
channels, one control channel (CCH) and six service channels (SCHs). In 2008,
the European Telecommunications Standards Institute (ETSI) allocated 30 MHz
of spectrum in the 5.9 GHz band for ITS. In 2014, the United States (U.S.)
National Highway Traffic Safety Administration (NHTSA) announced that it had
been working with the U.S. department of transportation on regulations that would
eventually mandate vehicular communication capabilities in new light vehicles by
2017 [4]. An envisioned VANET will consist of (1) vehicles with on-board sensing
and transmitting units which form the network nodes; (2) stationary road side units
(RSUs) deployed on the sides of roads and connected to the Internet; and (3) a
set of wireless channels from the DSRC spectrum. An illustration of a VANET
infrastructure is shown in Fig. 1.1.

The embedded wireless communication capabilities will enable both vehicle-to-
vehicle (V2V) and vehicle-to-infrastructure (V2I[1]) communications. Many new ITS
applications will emerge with the support of V2V and V2I communications.
ITS applications include on-road safety and infotainment applications. Examples

---

[1]V2I communications refer to the bidirectional communications between an RSU and a vehicle.

© The Author(s) 2015

K. Abboud, W. Zhuang, *Mobility Modeling for Vehicular Communication Networks*,
SpringerBriefs in Electrical and Computer Engineering,
DOI 10.1007/978-3-319-25507-1_1

1

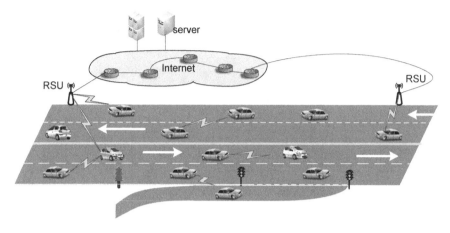

Fig. 1.1 Illustration of VANET infrastructure

of safety applications include emergency warning, lane changing assistance, and intersection coordination [5]. On the other hand, infotainment applications can provide (1) drivers with information about weather, maps, and directions to nearby petrol stations/restaurants, and (2) passengers with Internet access that includes web-surfing and multimedia applications [1, 3].

The implementation of VANET applications is dependent on the development of networking protocols that can guarantee reliable and efficient V2V and V2I communications. VANETs have specific characteristics that impose new challenges to the network development and operation when compared with traditional mobile ad hoc networks (MANETs). In traditional networks, nodes are either static or mobile with low speeds and independent movements. On the other hand, nodes in VANETs move with very high speeds, causing network fragmentations and rapid changes in the network topology [6]. The movement of vehicular nodes is also dependent on driver behaviors and the interaction with neighboring vehicles. Additionally, VANETs are susceptible to a large number of nodes, traffic[2] jams, and traffic density variations from time-to-time and from point-to-point on the same roads. Therefore, the networking protocols for VANETs should be robust to support such high node mobility.

1.2   Vehicle Mobility Models

Vehicle mobility is a major component of the VANET structure. The impact of node mobility on network topology dynamics and, consequently, on network protocol performance should be considered. Despite its high speed and randomness, the

---

[2]The term traffic refers to vehicle traffic in this brief.

Table 1.1 Traffic flow state for different vehicle densities [7]

Density (veh/km/lane)	Traffic flow condition
0–26	Uncongested flow conditions (low)
26–42	Near-capacity flow conditions (intermediate)
42–62	Congested flow conditions (high)

vehicle movement is restricted by road topology, speed limits, traffic rules, and movement of nearby vehicles. Therefore, based on these metrics, vehicle movement follows certain patterns [7]. Modeling vehicle traffic characteristics has attracted great attention from researchers in transportation engineering for many years. From the various vehicle traffic characteristics, the distance headway and its variations over time play an essential role in changing the network topology. The distance headway (or the intervehicle distance) is the distance between identical points on two consecutive vehicles on the same lane. In general, vehicle mobility models in the literature can be categorized into three (microscopic, mesoscopic, and macroscopic) types according to the detail level of the interactions among vehicles that the model characterizes [7–9]. A macroscopic distance headway model describes the average distance headway over a highway. The average distance headway is equal to reciprocal of the vehicle density. The vehicle density (D) is defined as the average number of vehicles per lane occupying a segment of the roadway [7]. The value of the vehicle density defines the traffic flow condition, an important metric for modelling vehicular traffic flow. Table 1.1 lists the traffic flow conditions corresponding to different traffic density ranges.

Mesoscopic traffic flow models describe the traffic flow with more details than macroscopic models. On the mesoscopic level, the behaviors of individual vehicles are characterized independently [7, 10]. A mesoscopic mobility model describes the distance headways of individual vehicles by independent and identically distributed random variables [7, 11]. That is, on the mesoscopic level, the traffic appears as a snap shot over the considered road segment. Different mesoscopic models have been proposed for different traffic flow conditions [7].

A microscopic model specifies time variations of a distance headway according to the driver behaviors and interactions with neighboring vehicles [7, 8]. On the microscopic level, the details of individual vehicle behaviors are modeled. The level of details include vehicular behaviors resulting from interacting with nearby vehicles. These behaviors include accelerating, decelerating, reacting to slowing leading vehicles, decisions on changing lanes, overtaking vehicles, etc. There are mainly two types of microscopic traffic flow models: the car following models and the cellular automata (CA) models. Car following models basically revolve around one simple rule: *keeping a safe distance ahead.*

Despite the accuracy of modeling the following-behavior of vehicles, there are several factors that limit the generality of car following models [12]. At a long distance headway, the interaction between vehicles is very low. In this case, vehicle behavior is not affected by the leading vehicle and will be in a free-driving mode which is not captured by a car-following model. Even in a highly dense situation, when the distance headway is small, a following vehicle may desire to move at a lower speed from the leading vehicle, and hence will not be in the following mode [12]. While car-following models are continuous models, CA-models are discrete. A CA-model describes a single lane road as a lattice of a number of equal sized cells [13]. The model includes a set of event-driven rules that define how a vehicle on road changes its speed and acceleration to traverse through cells (i.e., road segments). Some microscopic models has extended the scope of modeling to include the behavior of individual vehicles in response to the action of a group of leading vehicles, so called multi-anticipative car following model [14].

In order to characterize the spatiotemporal variations in VANET topology, a model that captures the time variations of distance headways need to be considered. Deterministic microscopic models (e.g. GM car following models [7]) do not reflect the realistic randomness in driver behavior. The two main factors that affect changes of a distance headway over time, i.e., the driver behaviors and interactions with neighboring vehicles, are both random. Furthermore, the correlation between a distance headway and its changes over a time period is not captured in a mesoscopic model. Therefore, to accurately model the temporal variations in VANET topology, a microscopic mobility model should be used. As pointed out earlier, microscopic mobility models in the literature include a set of deterministic and/or probabilistic rules that define how a vehicle on road changes its speed and/or acceleration in reaction to its neighboring vehicles' behaviors [7]. As such a model depends on the behaviors of neighboring vehicles over time, the analysis tends to take the form of case studies (e.g., [11]).

1.2.1   Vehicle Mobility Models for VANETs

In order to characterize the spatiotemporal variations in VANET topology, a realistic mobility model should be adopted. The vehicle mobility model development has advanced greatly to a highly sophisticated level that provides a close-to-real mimic of vehicle movements on roads. These models provide a base for vehicle traffic simulators. For example the use of Wiedemann's psycho-physical car-following model in the VISSIM microscopic vehicle traffic simulator [15]. Wiedemann is a psycho-physical car-following model that describes behaviors of individual vehicles according to their interactions with neighboring vehicles, their desired relative speeds, their relative positions, and some driver-dependant behaviors. The Wiedemann model accounts for four different driving modes: free driving, approaching, following, and breaking [16]. Although a realistic mobility model is desired to capture the spatiotemporal variations in VANET topology, the fine details of different characteristics of vehicle mobility (e.g., speed, acceleration, distance headway, driver psychological state ...etc.) may be unnecessary for VANET analysis. On the contrary, a sophisticated mobility model can complicate and further hinder the analysis.

In the VANET literature, there has been efforts to develop simple vehicle mobility models that facilitate VANET analysis while accurately describing vehicle mobility on roads [17, 18]. A vehicle mobility model should describe vehicle behavior on roads in enough details that account for its dependence on road topology, neighboring vehicles behaviors, and vehicle traffic condition. At the same time, the model granularity should not be too fine that prevents traceable VANET analysis. It is then useful to develop vehicle mobility models that can focus on the characteristics that are relevant to the system model in the considered VANET. For example, if a highway scenario is considered, the vehicle mobility model needs not to account for vehicle's behavior on signalized intersections or roundabouts. When considering a purely ad hoc vehicular network without RSUs, only the relative mobility between vehicles is needed to assess V2V communication analysis. On the other hand, when V2I communications are considered, the vehicle mobility model should be able to describe vehicle movement with respect to the road, i.e., how a vehicle traverses a specific road segment. In this case, cellular automata models are more suitable than car following models. In the following, a review of a few examples in the VANET literature, that propose different vehicle mobility models to facilitate specific VANET analysis, is presented. A zone-based mobility model that describes how a vehicle traverses RSU coverage area on a highway is proposed in [18]. The highway segment is divided into multiple spatial zones and a vehicle moves from one zone to another according to a continuous-time Markov chain as illustrated in Fig. 1.2a. A vehicle remains inside a zone for a geometrically distributed time duration with an average time proportional to the average vehicle speed and inverse proportional to the zone length. Such a model spares the unnecessary details in inter-vehicle interaction, which are of lower influence on the overall vehicle movement when compared to the high relative speed between vehicles and RSUs [18]. An area-based mobility model is proposed for a city scenario in [19]. The street map of a city is split into smaller areas each containing at least one intersection, which constitute the states of a Markov process, as shown in Fig. 1.2b. Vehicles enter the system (city) according to a Poisson

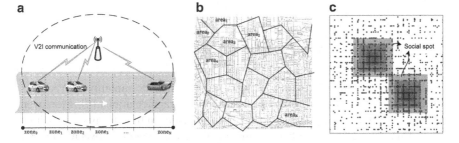

Fig. 1.2 Illustration of (**a**) zone-based vehicle mobility model proposed for V2I communication analysis in a highway scenario [18] (**b**) area-based vehicle mobility model for a city scenario [19], and (**c**) social-proximity vehicle mobility model for a city scenario (*blue dots* indicate vehicles) [20]

process. Once in an area, a vehicle remains in it for a time interval and then either jumps to another area according to a Markov process or leaves the system. Using such model, important macroscopic traffic characteristics can be modeled (e.g., vehicle density in each area) without the complexity of intervehicle interactions [19]. A social-proximity-based mobility model is proposed in [20]. It assumes that vehicle movement is restricted by a fixed social spot consisting of multiple tiers and defined around a city block. The model considers a city grid with a number of social spots uniformly distributed in the city. Each vehicle uniformly picks its social spot, which defines its mobility region throughout the network run-time. The spatial distribution of vehicles decays from the central social spot towards the border of the mobility region according to a power-law distribution. Figure 1.2c illustrates a network with two social spots. This model facilitates capacity-delay analysis in VANETs, by sparing the details of intervehicle correlations [20]. Vehicle speed variations over time are modeled in [17, 21]. The time is split into exponentially distributed time intervals, so called *mobility intervals*. At each mobility interval, a vehicle chooses its speed randomly and independently of neighboring vehicle's speeds and of its speed at the previous mobility interval. Vehicle speed remains constant during a single mobility interval. Since the model assumes that individual vehicle speed is temporally independent and is independent of neighboring vehicles' speeds, the model can be viewed as an extension of a mesoscopic model (uniformly distributed speeds).

1.3   Vehicle Mobility Effect on VANETs

Due to high and variable relative speeds, a VANET is subject to frequent spatiotemporal variations in its topology. Network topology changes and fragmentations can be measured by the switching of communication links between connection and disconnection. Another measure of network topology variations is the change in vehicle's local neighborhood. That is, the change in the set of nodes that are within the transmission range of a vehicle, i.e., vehicle's neighbors (VNs).

1.3.1   Communication Link Duration

Unlike traditional mobile ad hoc networks, the high node mobility in VANETs can cause frequent network topology changes and fragmentations [6]. Moreover, VANETs are susceptible to vehicle density variations from time-to-time throughout a day. This imposes new challenges in maintaining connections between vehicular nodes. The length of the communication link and its lifetime between network nodes are critical issues that determine the performance of network protocols. From a routing protocol perspective, the communication link length, i.e., the hop length, determines the route length between source and destination pairs [22, 23]. The route length is an important metric in shortest path algorithms, which are a base of many routing protocols. Additionally, a short communication link lifetime can disrupt an on-going packet transmission between two nodes, thus triggering a new route discovery procedure. The interruptions of information transmissions not only lead to a reception failure, but can also result in an increase in the control packet overhead associated with topology updates and route discovery/maintenance processes. Hence, short communication link lifetimes can inflict wastage of the limited radio bandwidth [24]. As a result, vehicle mobility poses a problem that needs to be addressed in designing routing protocols for VANETs [25–27]. From a medium access control (MAC) perspective, the communication link length determines the amount of channel spatial reuse and the amount of channel access that the nodes within a hop have. Simulations of the IEEE 802.11p standard MAC protocol have shown that a large relative speed between nodes (i.e., short link lifetime) reduces the channel access time of vehicular nodes [28]. Different MAC solutions that account for vehicle mobility have been proposed for VANETs [18, 29–32]. The diverse potential applications for VANETs vary in their connection time requirements. For example, safety applications do no require a long link lifetime when compared to multimedia applications [3]. Hence, analyzing the communication link length and lifetime is essential for designing efficient MAC protocols that support different application requirements in VANETs [28, 33].

1.3.2 Vehicle's Neighborhood

Inflicted by vehicle mobility, the fluctuations of communication links, between connections and disconnections, are accompanied by vehicles moving in and out of the communication range of other vehicles resulting in changes in vehicles one-hop neighborhood. Frequent changes in a vehicle's neighborhood consume network radio resources and cause service disruption for the network protocols (e.g., resource allocation in centralized MAC [32, 34], route discovery [27], message delivery, node cluster stability [35–37]). Characterizing the impact of vehicle mobility on vehicle's one-hop neighborhood is crucial for the design of network protocols, especially for safety applications. Most of the safety applications depend on vehicles broadcasting safety messages to all the nodes within their communication ranges, i.e., in their one-hop neighborhood [5].

In addition to vehicle's one-hop neighborhood, vehicle's two-hop neighborhood is also important to characterize the network topology.[3] Vehicle's two-hop neighborhood plays a crucial role in network protocols. When a source vehicle broadcasts a packet, a MAC protocol should be designed such that no other vehicle within

---

[3]The set of nodes between two-hop vehicles is referred to as vehicle's two-hop neighborhood, i.e., nodes in two-hop neighborhood of a reference vehicle are at most two-hop away from it, which includes its VNs.

the source vehicle's two-hop neighborhood transmits at the same time, otherwise the so-called hidden node problem can occur [38]. In multi-hop routing protocols, two-hop vehicles' neighborhood determines the availability and strength of multi-hop connections. The communication ranges of vehicles that are two-hop away overlap and share common VNs. A common VN can be made responsible to relay packets between the two-hop vehicles. The overlapping between the vehicles' coverage ranges can increase, due to vehicle mobility, until the vehicles become one-hop neighbors. This can consume radio resources due to redundant forwarding, rediscovery of routes, and packet collisions [30]. When the distance between two-hop vehicles increases, the overlapping between vehicles' coverage ranges diminishes until the two-hop connection between the vehicles breaks. This can result in disruption to on-going transmissions, which may trigger route discovery processes [39]. It is necessary to characterise the impact of vehicle mobility on increasing and decreasing the overlapping between the coverage ranges of two-hop vehicles. Such characterization should account for the change in the number of common VNs between the two-hop vehicles.

Despite the importance of characterizing spatiotemporal variations in VANET topology as a measure of node mobility impact on network structure, this has taken the form of empirical studies [6] or case studies [11] in the literature.

1.4   Brief Objectives and Outline

Due to the lack of mathematical analysis of spatiotemporal variations in VANET topology and the unavailability of microscopic mobility model that facilitates such analysis, as discussed in previous sections, this brief introduces a novel stochastic microscopic vehicle mobility model and presents the use of this model in characterizing spatiotemporal changes in VANET topology. The rest of this brief is organized as follows. Chapter 2 introduces the system model under consideration. Chapter 3 presents the stochastic microscopic mobility model that describes the time variation of individual distance headway on a single lane highway, which is consistent with highway data patterns from empirical and simulated data sets [40, 41]. In addition, the mobility model is extended to a group mobility model that describes the temporal variations in a system of consecutive distance headways [36, 37]. Chapter 4 introduces a stochastic analysis of spatiotemporal variations in VANET topology in terms of communication link duration and the vehicles' neighborhood [36, 37, 40, 41]. Finally, Chap. 5 concludes this brief and identifies some further research topics.

References

1. N. Lu, N. Cheng, N. Zhang, X. Shen, and J. W. Mark, "Connected vehicles: Solutions and challenges," *IEEE J. Internet of Things*, vol. 1, no. 4, pp. 289–299, 2014.
2. H. A. Omar, N. Lu, and W. Zhuang, "Wireless access technologies for vehicular network safety applications," *IEEE Network*, to appear.
3. H. T. Cheng, H. Shan, and W. Zhuang, "Infotainment and road safety service support in vehicular networking: From a communication perspective," *Mechanical Systems and Signal Processing*, vol. 25, no. 6, pp. 2020–2038, 2011.
4. J. Harding, G. Powell *et al.*, "Vehicle-to-vehicle communications: Readiness of V2V technology for application," U.S. Department of Transportation, Tech. Rep. DOT HS 812 014, 2014.
5. "Vehicle safety communications project: task 3 final report: identify intelligent vehicle safety applications enabled by DSRC," CAMP Vehicle Safety Communications Consortium, National Highway Traffic Safety Administration (NHSTA), U.S. Department of Transportation, Tech. Rep. DOT HS 809 859, 2005.
6. F. Bai and B. Krishnamachari, "Spatio-temporal variations of vehicle traffic in VANETs: facts and implications," in *ACM Proc. Int. workshop on Vehicular InterNetworking*, 2009, pp. 43–52.
7. A. May, *Traffic Flow Fundamentals*. Prentice Hall, 1990.
8. M. Krbalek and K. Kittanova, "Theoretical predictions for vehicular headways and their clusters," *Physics: Data Analysis, Statistics and Probability (arXiv)*, 2012.
9. L. Li, W. Fa, J. Rui, H. Jian-Ming, and J. Yan, "A new car-following model yielding log-normal type headways distributions," *Chinese Physics B*, vol. 19, no. 2, 2010.
10. S. Hoogendoorn and P. Bovy, "State-of-the-art of vehicular traffic flow modelling," *Proceedings of the Institution of Mechanical Engineers, Part I: Journal of Systems and Control Engineering*, vol. 215, no. 4, pp. 283–303, 2001.
11. G. Yan and S. Olariu, "A probabilistic analysis of link duration in vehicular ad hoc networks," *IEEE Trans. Intelligent Transportation Systems*, vol. 12, no. 4, pp. 1227–1236, 2011.
12. R. Luttinen, "Statistical properties of vehicle time headways," *Transportation Research Record*, no. 1365, 1992.
13. K. Nagel and M. Schreckenberg, "A cellular automaton model for freeway traffic," *Journal de Physique I*, vol. 2, no. 12, pp. 2221–2229, 1992.
14. H. Lenz, C. Wagner, and R. Sollacher, "Multi-anticipative car-following model," *European Physical Journal B-Condensed Matter and Complex Systems*, vol. 7, no. 2, pp. 331–335, 1999.
15. PTV, "VISSIM 5.40 user manual," *Karlsruhe, Germany*, 2012.
16. M. Fellendorf and P. Vortisch, "Microscopic traffic flow simulator VISSIM," *Fundamentals of Traffic Simulation*, pp. 63–93, 2010.
17. K. A. Hafeez, L. Zhao, B. Ma, and J. W. Mark, "Performance analysis and enhancement of the DSRC for VANET's safety applications," *IEEE Trans. Vehicular Technology*, vol. 62, no. 7, pp. 3069–3083, 2013.
18. T. H. Luan, X. Ling, and X. Shen, "MAC in motion: impact of mobility on the MAC of drive-thru internet," *IEEE Trans. Mobile Computing*, vol. 11, no. 2, pp. 305–319, 2012.
19. Y. Li, D. Jin, Z. Wang, P. Hui, L. Zeng, and S. Chen, "A markov jump process model for urban vehicular mobility: modeling and applications," *IEEE Trans. Mobile Computing*, vol. 13, no. 9, pp. 1911–1926, 2014.
20. N. Lu, T. H. Luan, M. Wang, X. Shen, and F. Bai, "Bounds of asymptotic performance limits of social-proximity vehicular networks," *IEEE/ACM Trans. Networking*, vol. 22, no. 3, pp. 812–825, 2014.
21. K. A. Hafeez, L. Zhao, Z. Liao, and B. N.-W. Ma, "Impact of mobility on VANETs' safety applications," in *Proc. IEEE Globecom*, 2010, pp. 1–5.
22. K. Abboud and W. Zhuang, "Impact of node clustering on routing overhead in wireless networks," in *Proc. IEEE Globecom*, 2011, pp. 1–5.

23. N. Wisitpongphan, F. Bai, P. Mudalige, V. Sadekar, and O. Tonguz, "Routing in sparse vehicular ad hoc wireless networks," *IEEE J. Selected Areas in Communications (JSAC)*, vol. 25, no. 8, pp. 1538–1556, 2007.

24. F. Bai, N. Sadagopan, and A. Helmy, "BRICS: A building-block approach for analyzing routing protocols in ad hoc networks-a case study of reactive routing protocols," in *Proc. IEEE ICC*, vol. 6, 2004, pp. 3618–3622.

25. X. Wu, H. R. Sadjadpour, and J. Garcia-Luna-Aceves, "Routing overhead as a function of node mobility: modeling framework and implications on proactive routing," in *IEEE Proc. MASS*, 2007, pp. 1–9.

26. N. Wisitpongphan, F. Bai, P. Mudalige, and O. K. Tonguz, "On the routing problem in disconnected vehicular ad-hoc networks," in *IEEE. Proc. INFOCOM*, 2007, pp. 2291–2295.

27. F. Li and Y. Wang, "Routing in vehicular ad hoc networks: A survey," *IEEE Vehicular Technology Magazine*, vol. 2, no. 2, pp. 12–22, 2007.

28. W. Alasmary and W. Zhuang, "Mobility impact in IEEE802.11p infrastructureless vehicular networks," *Ad Hoc Networks*, vol. 10, no. 2, pp. 222–230, 2012.

29. M. Wang, Q. Shen, R. Zhang, H. Liang, and X. Shen, "Vehicle-density-based adaptive MAC for high throughput in drive-thru networks," *IEEE J. Internet of Things*, vol. 1, no. 6, pp. 533–543, 2014.

30. H. Omar, W. Zhuang, and L. Li, "VeMAC: A TDMA-based MAC protocol for reliable broadcast in VANETs," *IEEE Trans. Mobile Computing*, vol. 12, no. 9, pp. 1724–1736, 2013.

31. H. Zhou, B. Liu, T. H. Luan, F. Hou, L. Gui, Y. Li, Q. Yu, and X. Shen, "Chaincluster: Engineering a cooperative content distribution framework for highway vehicular communications," *IEEE Trans. Intelligent Transportation Systems*, vol. 15, no. 6, pp. 2644–2657, 2014.

32. H. Su and X. Zhang, "Clustering-based multichannel MAC protocols for QoS provisionings over vehicular ad hoc networks," *IEEE Trans. Vehicular Technology*, vol. 56, no. 6, pp. 3309–3323, 2007.

33. D. Kumar, A. A. Kherani, and E. Altman, "Route lifetime based optimal hop selection in VANETs on highway: an analytical viewpoint," *Networking Technologies, Services, and Protocols, Performance of Computer and Communication Networks, Mobile and Wireless Communications Systems*, pp. 799–814, 2006.

34. N. Chandra Rathore, R. Tomar, S. Verma, and G. Tomar, "CMAC: A cluster based MAC protocol for VANETs," in *Proc. IEEE CISIM*, 2010, pp. 563–568.

35. E. Souza, I. Nikolaidis, and P. Gburzynski, "A new aggregate local mobility (ALM) clustering algorithm for VANETs," in *Proc. IEEE ICC*, 2010, pp. 1–5.

36. K. Abboud and W. Zhuang, "Stochastic modeling of single-hop cluster stability in vehicular ad hoc networks," *IEEE Trans. Vehicular Technology*, 2015 (to appear).

37. K. Abboud and W. Zhuang, "Impact of node mobility on single-hop cluster overlap in vehicular ad hoc networks," in *Proc. ACM MSWiM*, 2014, pp. 65–72.

38. F. Borgonovo, A. Capone, M. Cesana, and L. Fratta, "ADHOC MAC: new MAC architecture for ad hoc networks providing efficient and reliable point-to-point and broadcast services," *Wireless Networks*, vol. 10, no. 4, pp. 359–366, 2004.

39. K. Abboud and W. Zhuang, "Impact of microscopic vehicle mobility on cluster-based routing overhead in VANETs," *IEEE Trans. Vehicular Technology, connected vehicle series*, 2015 (to appear).

40. K. Abboud and W. Zhuang, "Analysis of communication link lifetime using stochastic microscopic vehicular mobility model," in *Proc. IEEE Globecom*, 2013, pp. 383–388.

41. K. Abboud and W. Zhuang, "Stochastic analysis of single-hop communication link in vehicular ad hoc networks," *IEEE Trans. Intelligent Transportation Systems*, vol. 15, no. 5, pp. 2297–2307, 2014.

Chapter 2
System Model

2.1 VANET Scenario

Consider a connected VANET on a multi-lane highway with no on or off ramps. This brief focuses on a single lane with lane changes implicitly captured in the adopted mobility model. A single lane from a multi-lane highway is chosen instead of a single-lane highway, in order to be more realistic in a highway scenario. A vehicle can overtake a slower leading vehicle, if possible, and accelerate towards its desired speed.[1] Assume that the highway is in a steady traffic flow condition defined by a time-invariant vehicle density. Let D denote the vehicle density in vehicle per kilometer. Three levels of D are considered: low, intermediate, and high vehicle densities as in Table 1.1. However, in this research the case when the vehicle density is changing among the three levels is not considered. Additionally, this work does not consider the case of increasing/decreasing vehicle density within the same level of density. The system model focuses only on a single direction traffic flow. All the vehicles have the same transmission range, denoted by R. Any two nodes at a distance less than R from each other are one hop neighbors. The set of vehicles, that are within the coverage range R of a vehicle, is referred to as *vehicle's one-hop neighborhood* as illustrated in Fig. 2.1a. The length of a hop is defined as the distance to the furthest node within the transmission range of a reference node, which is upper bounded by R as illustrated in Fig. 2.1b. The furthest node within the transmission range of a reference vehicle is referred to as *hop edge node*. Let H denote the hop length with respect to a reference node. Assume that the transmission range is much larger than the width of the highway such that a node can communicate with any node within a longitudinal distance

---

[1]In a single-lane highway, the vehicle traffic gradually converges into a number of platoons lead by the slower vehicles on the highway [1].

© The Author(s) 2015
K. Abboud, W. Zhuang, *Mobility Modeling for Vehicular Communication Networks*,
SpringerBriefs in Electrical and Computer Engineering,
DOI 10.1007/978-3-319-25507-1_2

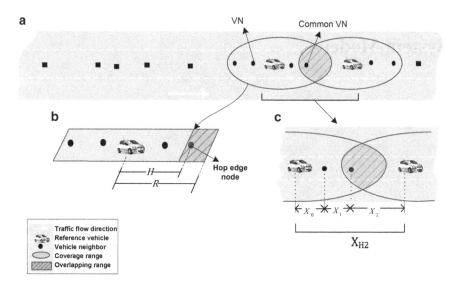

Fig. 2.1 Illustration of the system model under consideration

of R from it.[2] Time is partitioned with a constant step size. Let X_i be the distance headway between node i and node $i+1$, $i = 0, 1, 2, \ldots$. The distance headway is the distance between two identical points on two consecutive vehicles on the same lane. Define $X_i = \{X_i(m), m = 0, 1, 2 \ldots\}$ to be a discrete-time stochastic process of the ith distance headway, where $X_i(m)$ is a random variable representing the distance headway of node i at the mth time step. At any time step, $X_i(m) \in [\alpha, X_{\max}]$ for all $i, m \geq 0$, where α and $X_{\max}$ is the minimum and maximum distance headway, respectively. Furthermore, assume that the distance headways (X_i for all $i \geq 0$) are independent with identical statistical behaviors. For notation simplicity, the index i from X_i is omitted when referring to an arbitrary distance headway. In this analysis, the 0th time step refers to the time when the network has just established. A two-hop neighborhood between two reference vehicles is the set of vehicles between two reference vehicles that are connected via two-hop connection. Let $\mathbb{X}_{H2}$ denote the sequence of distance headways between two reference vehicles that are two-hop apart as illustrated in Fig. 2.1c. The set of nodes between the two-hop vehicles is referred to as *vehicle's two-hop neighborhood*. The vehicles are assumed to be distributed on the highway according to a stationary probability distribution of the distance headways when the network is first established. Let μ and σ be the mean and the standard deviation of the distance headway in meters, respectively, where $\mu = 1000/D$ and σ are constant system parameters

---

[2]Typically, the transmission range covers a circular area with a radius R centred at the node. However, since the transmission range is much larger than the width of the road, the area covered by the transmission range can be approximated by a rectangular area with length $2R$.

and take different values according to the vehicle density. Throughout this brief, $F_Y(y), P_Y(y), f_Y(y), Q_Y(y)$, and $E[Y]$ are used to denote the cumulative distribution function (cdf), the probability mass function (pmf), the probability density function (pdf), the probability generating function, and the expectation of random variable Y, respectively.

2.2 Node Mobility

This research focuses on the distance headway and the spatial distribution of vehicles along the road. Therefore, the adopted vehicular mobility model is used to describe the distance headway. Unless otherwise mentioned, on a macroscopic level, three different traffic flow conditions are studied separately: uncongested, near-capacity, and congested. Each traffic flow condition corresponds to a range of vehicle densities according to Table 1.1 [2]. The uncongested, near capacity, and congested traffic flow conditions correspond to low, intermediate, and high vehicle densities, respectively. Furthermore, each traffic flow condition corresponds to a unique microscopic and a unique mesoscopic distance headway model. On a microscopic level, the time variations of the distance headway is modeled by a discrete-time finite-state Markov chain. Details of the microscopic model are given in Chap. 3.

2.2.1 Mesoscopic Mobility Model

The literature of mesoscopic models focuses on the time-headway, which is the elapsed time of the passage of identical points on two consecutive vehicles [2]. For an uncongested traffic flow condition, the exponential distribution has been shown to be a good approximation for the time headway distribution [2]. With a low vehicle density, interactions between vehicles are very low and almost negligible. As a result, vehicles move independently at a maximum speed [2]. It is reasonable to assume that, over a short time interval of interest, vehicles move at constant velocity and do not interact with each other [3, 4]. Therefore, for a low vehicle density, it is assumed that the distance headway has the same distribution as the time headway with parameters properly scaled. The distance headways at any time step are independent and identically distributed (i.i.d.) with an exponential probability density function (pdf)

$$f_{X_i}(x) = \frac{1}{\mu}e^{-\frac{x}{\mu}}, \quad x \geq 0. \tag{2.1}$$

In this case, the mean and the standard deviation of the distance headway are equal $(\mu = \sigma = \frac{1000}{D})$. According to the distribution, $P(X_i \leq \alpha) > 0$; however, for simplicity, the effect of this probability is ignored.[3]

In the literature, the Gaussian distribution is used to model the time headway for a congested traffic flow condition [2]. Although the time headway is almost constant for a high vehicle density, driver behaviors cause the time headway to vary around that constant value. Therefore, the Gaussian distribution model for the time headway characterizes the driver attempt to drive at a constant time headway [2]. With the same argument, a distance headway is assumed to vary around a constant value with a Gaussian distribution. The pdf of the distance headway is approximately given by

$$f_{X_i}(x) = \frac{1}{\sqrt{2\pi}\sigma}e^{-\frac{(x-\mu)^2}{2\sigma^2}}, \quad x \geq 0. \tag{2.2}$$

The standard deviation σ for a high vehicle density is given by[4] $\sigma = \frac{(\mu-\alpha)}{2}$.

For a near-capacity traffic flow condition, empirical pdfs for inter-vehicle distances show that neither an exponential nor a Gaussian distribution is a good fit [5]. Hence, the inter-vehicle distances are assumed to follow a general distribution, Pearson type III, that was originally proposed for time headways [2]. With an intermediate vehicle density, the pdf of the distance headway is approximately given by

$$f_{X_i}(x) = \frac{\lambda^z}{\Gamma(z)}(x - \alpha)^{z-1}e^{-\lambda(x-\alpha)}, \quad x \geq \alpha \tag{2.3}$$

where λ and z are the scale and shape parameters of the general Pearson type III distribution, respectively, and $\Gamma(z) = \int_0^\infty u^{z-1}e^{-u}du$ is the gamma function. The parameters λ and z are related to μ and σ according to the following relations [2]

$$\lambda = \frac{\mu - \alpha}{\sigma^2}, \quad z = \frac{(\mu - \alpha)^2}{\sigma^2}. \tag{2.4}$$

---

[3]$P(X_i \leq \alpha) = 1 - e^{-D\alpha}$. For example, for $D = 6$ veh/km and $\alpha = 6.7$ m [2], $P(X_i \leq \alpha) = 0.04$. The probability $P(X_i \leq \alpha)$ increases with D.

[4]The guidelines used for calculating the variance of time headway are given in [2]. With $\sigma = \frac{(\mu-\alpha)}{2}$, $P(X_i > \alpha) = 0.977$ [2]. For a congested traffic flow condition (i.e., $D \geq 42$ veh/km) and $\alpha = 6.7\,m$ [2], $P(X_i \leq 0) \leq 2.8 \times 10^{-3}$.

References

1. L. C. Edie and R. S. Foote, "Traffic flow in tunnels," in *Highway Research Board Proceedings*, 1958.
2. A. May, *Traffic flow fundamentals*. Prentice Hall, 1990.
3. P. Izadpanah, "Freeway travel time prediction using data from mobile probes," Ph.D. dissertation, University of Waterloo, 2010.
4. F. Hall, "Traffic stream characteristics," *Traffic Flow Theory. US Federal Highway Administration*, 1996.
5. X. Chen, L. Li, and Y. Zhang, "A markov model for headway/spacing distribution of road traffic," *IEEE Trans. Intelligent Transportation Systems*, vol. 11, no. 4, pp. 773–785, 2010.

Chapter 3
Microscopic Vehicle Mobility Model

Unlike traditional mobile ad hoc networks, the high node mobility in VANETs can cause frequent network topology changes and fragmentations. As discussed in Sect. 1.3, any change in network topology is directly or indirectly related to the change in distance headways among vehicles. This chapter presents a novel microscopic mobility model to facilitate VANET analysis. Firstly, A discrete-time finite-state Markov chain with state dependent transition probabilities is proposed to model the distance headway. The model captures the time variations of a distance headway and its dependency on distance headway value. Secondly, highway vehicular traffic is simulated using microscopic vehicle traffic simulator, VISSIM. Vehicle trajectory data collected from highways in the U.S. and that simulated by VISSIM are used to demonstrate the validity of the proposed mobility model for three vehicle density ranges. Finally, the proposed mobility model is extended to a group mobility model that describes the time variations of a system of distance headways between two non-consecutive vehicles. Using lumpability theory, a Markov chain with reduced state-space is proposed to represent the mobility of a group of vehicles.

3.1 Individual Mobility Model

The stochastic process, X_i, is modeled as a discrete-time finite-state Markov chain, inspired by Chen et al. [1] and Gong et al. [2]. The Markov chain, illustrated in Fig. 3.1, has N_{max} states corresponding to N_{max} ranges of a distance headway. The length of the range covered by each state is a constant, denoted by L_s in meters. The jth state covers the range $[x_j, x_j + L_s)$, $0 \leq j \leq N_{max} - 1$, where $x_j = \alpha + jL_s$. At any time step, $X_i(m) = x_j$ denotes that the distance headway X_i is in the jth state, for all $i, m \geq 0$, and $0 \leq j \leq N_{max} - 1$. Let $N_R = \frac{R-\alpha}{L_s}$ be the integer number of states that cover distance headways within R. Hence, the states with indices

© The Author(s) 2015
K. Abboud, W. Zhuang, *Mobility Modeling for Vehicular Communication Networks*,
SpringerBriefs in Electrical and Computer Engineering,
DOI 10.1007/978-3-319-25507-1_3

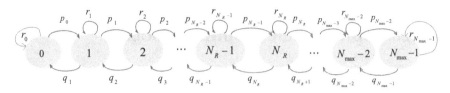

Fig. 3.1 Illustration of the proposed discrete-time $N_{\max}$-state Markov chain model of the distance headway

$j \in \{0, 1, 2, \ldots, N_R - 1, N_R, N_R + 1, \ldots, N_{\max} - 2$, and $N_{\max} - 1\}$ correspond to the quantized distances $x_j \in \{\alpha, \alpha + L_s, \alpha + 2L_s, \alpha + 3L_s, \ldots, \alpha + (N_R - 1)L_s, R, \alpha + (N_R + 1)L_s, \ldots, X_{\max} - L_s$, and $X_{\max}\}$, respectively. Within a time step, a distance headway in state j can transit to the next state, the previous state, or remain in the same state with probabilities p_j, q_j, or r_j, $0 \le j \le N_{\max} - 1$, respectively. Without loss of generality, assume that probability transition matrix is a positive-definite tri-diagonal and is given by Abboud and Zhuang [3, 4]

$$
M = \begin{pmatrix}
r_0 & p_0 & 0 & \cdots & & \cdots & 0 \\
q_1 & r_1 & p_1 & 0 & & \cdots & \vdots \\
0 & q_2 & r_2 & p_2 & 0 & & \vdots \\
\vdots & \ddots & \ddots & \ddots & & \ddots & 0 \\
0 & \cdots & 0 & q_{N_{\max}-2} & r_{N_{\max}-2} & p_{N_{\max}-2} \\
0 & \cdots & \cdots & & 0 & q_{N_{\max}-1} & r_{N_{\max}-1}
\end{pmatrix}.
\tag{3.1}
$$

The tri-diagonal structure of M is due to the fact that the values of a distance headway at consecutive time steps are highly correlated, for a short time step, such as $\tau \le \frac{L_s}{\bar{v}}$, where $\bar{v}$ is the maximum relative speed between vehicles.[1] The following state-dependent transition probability functions are proposed

$$
p_j = p\left(1 - \beta\left(1 - \frac{x_j}{X_{\max}}\right)\right)
$$

$$
q_j = q\left(1 - \beta\left(1 - \frac{x_j}{X_{\max}}\right)\right)
$$

$$
r_j = 1 - p_j - q_j, \qquad 0 \le j \le N_{\max} - 1, \, 0 \le p, q, \beta \le 1
\tag{3.2}
$$

---

[1]Consider an i.i.d. desired vehicle speed with a mean of 100 km per hour and a standard deviation of 10 km per hour, i.e., $P(\bar{v} \le 36) = 0.99$. In this case, the choice of $\tau = 2$ s for $L_s = 20$ m, reduces the transition probability of the distance headway to a non-neighboring state to less than 0.0054.

where p, q, and β are constants that depend on the vehicle density. For a low vehicle density, β is close to zero, and therefore the transition probabilities are independent of the state value, x_j. The value of β increases as the vehicle density increases, and thus increases the dependency on the state value. Equation (3.2) can be explained as follows. In a low vehicle density, distance headways are relatively large. Hence, a vehicle moves freely with a desired speed [5]. In such a scenario, the distance headway value does not affect the driver's choice to keep/change the speed, since the distance headway is large enough. On the other hand, in a high vehicle density situation, distance headways are relatively small. Hence, vehicles move with high constraints to keep a safe distance ahead. In such a scenario, the distance headway value has a high impact on the driver's behavior and his/her choice to keep/change the speed (and consequently the distance headway). The constant β is comparable with the driver strain constant used in [6], whereas $\beta = 0$ in [7]. The transition probabilities of the distance headway to neighboring states increase with the distance headway value, when $\beta > 0$ in (3.2). This is due to the fact that a larger distance headway results in less constraints in driving.

3.1.1 Model Validation

In order to verify the dependency of the distance headway transition probability on its current state, the transition probability matrix is computed using (1) empirical vehicle trajectory data collected from highways provided by Next Generation Simulation (NGSIM) community and available online [8], and (2) simulated vehicle trajectory data generated by VISSIM microscopic vehicle traffic simulator.

3.1.1.1 VISSIM Simulation Data

The vehicles in VISSIM simulator move according to Wiedemann's microscopic mobility model. Wiedemann is psycho-physical car-following model that describes behaviors of individual vehicles according to their interactions with neighboring vehicles, their desired relative speeds, their relative positions, and some driver-dependant behaviors. The Wiedemann model accounts for four different driving modes: free driving, approaching, following, and breaking [9]. The Wiedemann 99 model is adopted, which is designed for a highway scenario with its parameters set to the default values suggested in [10]. The VISSIM data set was obtained via six 30-min simulations of a three-lane highway traffic for different vehicle densities. The highway is a closed loop, and the vehicles enter the highway with a traffic flow of (3052.8, 1914.2, 854.6, and 683.7) vehicle per hour per lane for 1000 s, resulting in vehicle densities of (42, 26, 16, and 9), respectively. The VISSIM data points associated with vehicles entering the highway or changing lanes are not included in the analysis.

3.1.1.2 NGSIM Empirical Data

Two NGSIM data sets are used: *I-80-Main-Data* and *US-101-Main-Data*, which
were collected from a seven lane highway for a section of 500 and 640 m, respectively. From the NGSIM data sets, the data points associated with the following
vehicles are excluded: vehicles (1) on an on-ramp lane, (2) on an off-ramp lane,
(3) at the end of the section, or (4) undertaking a lane change. The NGSIM data sets
is only available with intermediate-to-high vehicle densities with $X_{max} = 100$ m.

To obtain the transition probabilities, the NGSIM and VISSIM data sets are
mapped into a sequence of quantized state values (x_j) with a predefined state length
L_s, where $x_j \in [0, X_{max}]$ in meters and $0 \le j \le N_{max} - 1$. For each state j, the
transition probabilities for the distance headway are determined by counting the
number of occurrences of each transition. Let $n_{j,j'}$ be the number of transitions of a
distance headway from state j to state $j', 0 \le j' \le N_{max} - 1$, within a time step of
length τ, and $n_j = \sum_{j'=0}^{N_{max}-1} n_{j,j'}$ be the number of time steps at which the distance
headway is in state j. The transition probability from state j to j' is calculated by
$p_{j,j'} = \frac{n_{j,j'}}{n_j}, 0 \le j, j' \le N_{max} - 1$.

Figure 3.2 plots the transition probabilities (and their standard deviation) from
state j to its direct neighboring states and to itself for different x_j values, with the
default data recording values: $L_s = 1$ meter and $\tau = 0.1$ s for the NGSIM data set,
and $L_s = 2$ m and $\tau = 0.2$ s for the VISSIM data set. The results show a dependency
of the transition probabilities on the x_j value. The weighted linear regression (LR) is
used to fit the transition probabilities in Fig. 3.2, with $n_{j,j'}$ being the weight of each
$p_{j,j'}$ data point. The transition probabilities p_j and q_j increase with the quantized
state value, x_j, which agrees with (3.2). The values of p, q, and β are calculated
according to the resulting weighted LR fit and are given in the figure legends. The
results show that $p_{j,j'}$ is smaller than 10^{-3} for $|j - j'| > 1$, and is therefore neglected,
which is consistent with the tri-diagonal transition matrix assumption given in (3.1).

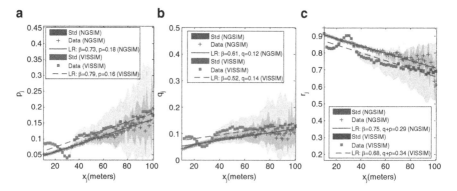

Fig. 3.2 The transition probability from state j to (**a**) state $j + 1$, (**b**) state $j - 1$, and (**c**) state j,
for different x_j values from NGSIM and VISSIM data for intermediate to high vehicle densities.
Results for the weighted LR fit model for (3.2) are given in the legends

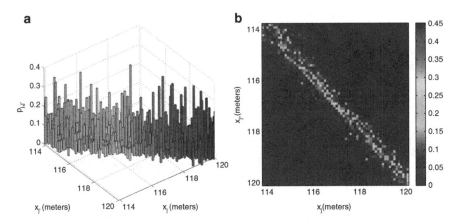

Fig. 3.3 Probability transition matrix for 100 quantized values of $x_j, x_{j'} \in [114, 120]$ with $L_s = 0.1$ m and $\tau = 0.1$ s. The matrix is calculated based on NGSIM data. The transition matrix is presented by (**a**) 3D plot and (**b**) color map plot

Figure 3.3 plots the transition probability matrix calculated from the NGSIM data for $x_j, x_{j'} \in [114, 120]$, $L_s = 0.1$ m, and $\tau = 0.1$ s. It is observed from Fig. 3.3 that, for a reduced value of the ratio L_s / τ, $p_{j,j'}$ increases for $|j - j'| > 1$. The reason behind this is that the inequality $(\tau \leq \frac{L_s}{v})$ is not satisfied. As a result, the value of transition probability to non-neighboring states is not negligible, as discussed in the previous section. Figure 3.4 plots the transition probability to the next state (and its standard deviation) for different x_j values and different vehicle densities. The results show that, the larger the vehicle densities, the higher the state-dependency of the transition probabilities. The resulting β values for different vehicle densities are plotted in Fig. 3.5, which shows an approximate linear relation between β and the vehicle density. This agrees with the proposed transition probability functions in (3.2).

It should be noted that the proposed microscopic model does not explicitly describe how and when lane changes occur nor does it describe impacts of lane-changes on the time variations of distance headways. However, the model implicitly captures the impact of lane changes on maintaining the ability of the vehicles to overtake slower vehicles and accelerate towards their desired speed. This is captured in the parameters p, q, and β which can be tuned from empirical/simulated multi-lane highway trajectory data for one of the lanes as done earlier in this section.

3.2 Group Mobility Model

Consider two non-consecutive vehicles, the relative mobility between them is effected not only by the mobility of the two vehicles, but also by the mobility of the vehicles between them. In this section, the mobility of a group of vehicles is

a   **b**

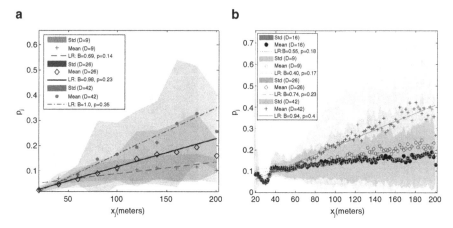

Fig. 3.4 Transition probability from state j to state $j + 1$, for different x_j values from VISSIM data for vehicle densities of (**a**) 9, 26, and 42 veh/km with $L_s = 20$ m and $\tau = 2$ s and (**b**) 9, 16, 26, and 42 veh/km, with $L_s = 2$ m and $\tau = 0.2$ s. Results of the weighted linear regression fit model for (3.2) are given in the legends

Fig. 3.5 State dependency parameter, β for different D values calculated based on VISSIM data

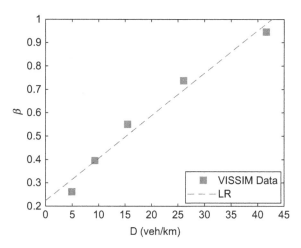

investigated, where every vehicle moves according to the individual mobility model described in the previous section. Consider two non-consecutive vehicles and let N be the number of distance headways separating them. That is, the two reference vehicles are separated by $N - 1$ vehicles. The relative mobility between the two reference vehicles is governed by the total distance between them. As the total distance decreases, the two vehicles approach each other causing their coverage ranges to overlap. On the other hand, as the distance between the two reference vehicle increases, the vehicles move further apart from each other. The distance between two non-consecutive vehicles and its variations over time play an essential role in determining the length and the duration of communication links between

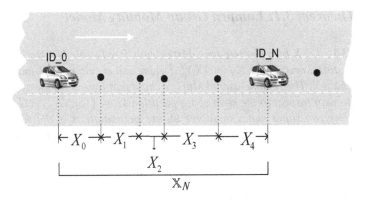

Fig. 3.6 Two non-consecutive vehicles with IDs 0 and N, separated by $N = 5$ distance headways and $\mathbb{X}_N = (X_0, X_1, X_2, X_3, X_4)$

network nodes, which are the bases of network topology. The distance between two non-consecutive vehicles is equal to the sum of the distance headways between the two nodes. Label the $(N+1)$ nodes with IDs $0, 1, \ldots, N$, where the following vehicle has ID 0 and the leading vehicle has ID N. For notation simplicity, let $\mathbb{X}_N = (X_i)_{i=0}^{N-1}$ be the sequence of N distance headways between the two reference vehicles as illustrated in Fig. 3.6, where $\mathbb{X}_N(m) = (X_i(m))_{i=0}^{N-1}$, and $\{\mathbb{X}_N(m) \in (s_0, s_1, \ldots, s_N)\} \equiv \{X_i(m) \in s_i, \forall i \in [0, N-1]\}$. Consider the total distance between the two reference vehicles at the 0th time step, i.e., $\sum_{i=0}^{N-1} X_i(0)$. The sequence of N i.i.d. distance headways is an N-dimensional Markov chain, where each headway, X_i, is a birth and death Markov chain as described in Sect. 3.1. For clarity, the term *state* refers to a state in the original Markov chain, X, the term *super state* refers to a state in the N-dimensional Markov chain, and the term *lumped state* refers to a set of super states (to be discussed later in this section). Additionally, parentheses ( ) are used for a sequence, while curly brackets { } are used for a set. A super state in the N-dimensional Markov chain is a sequence of size N, in which the ith element represents the state (in the one dimensional (1D)-Markov chain) that the ith distance headway belongs to. That is, a super state, $(s_0, s_1, \ldots, s_{N-1})$, means that distance headway X_i is in state $s_i \in [0, N_{\max} - 1]$. The sum of N distance headways representing the distance between the two reference vehicles can be calculated from the N-dimensional Markov chain. The state space size of the N-dimensional Markov chain is equal to $N_{\max}^N$, making it subject to the state-space explosion problem when N is large. However, since the focus is on the sum of the N distance headways, the state space can be reduced according to the following theorem [11, 12].

3.2.1   Theorem 3.1: Lumped Group Mobility Model

Theorem 3.1. *Let X be a discrete-time, birth-death, irreducible Markov chain with N_{max} finite states, and let set $\mathbb{X}_N = (X_i)_{i=0}^{N-1}$ represent a system of N independent copies of chain X. The N-dimensional Markov chain that represents the system, $\mathbb{X}_N$, is lumpable with respect to the state space partition $\Omega = \{\Omega_0, \Omega_1 \dots, \Omega_{N_L}\}$, such that (s.t.) any two super states in subset Ω_i are permutations of the same set of states $\forall i \in [0, N_L - 1]$, where $N_L = \frac{(N_{max}+N-1)!}{N!(N_{max}-1)!}$ is the state space size of the lumped Markov chain.*

3.2.1.1   Proof of Theorem 3.1 [11, 12]

Let $\ddot{M}_N = \{\ddot{M}_N(S_i, S_j)\}$, $0 \leq S_i, S_j \leq N_{max}^N - 1$, be the transition matrix of the N-dimensional Markov chain that represents the system of N independent copies of the one-dimensional Markov chain, X, with transition matrix $M = \{M(u_i, u_j)\}$, $0 \leq u_i, u_j \leq N_{max} - 1$. A discrete-time Markov chain with stochastic transition matrix $\ddot{M}_N$ is lumpable with respect to the partition Ω if and only if, for any subsets Ω_i and Ω_j in the partition, and for any super states S_1 and S_2 in subset Ω_i [13],

$$\sum_{S \in \Omega_j} \ddot{M}_N(S_1, S) = \sum_{S \in \Omega_j} \ddot{M}_N(S_2, S). \tag{3.3}$$

Consider the left hand side (LHS) of (3.3). Since X is a birth-death process, the super state $S_1 = (u_0, u_1, \dots, u_{N-1})$, $0 \leq u_i \leq N_{max} - 1$, can transit to any super state in set $\mathbf{A} = \{(u'_0, u'_1, \dots, u'_{N-1})\}$, where state $u'_i \in \{u_i - 1, u_i, u_i + 1\}$, i.e., $|\mathbf{A}| \leq 3^{N_{max}}$. Let subsets $\mathbf{A_i} = \mathbf{A} \cap \Omega_i$ and $\mathbf{A_j} = \mathbf{A} \cap \Omega_j$. Since $\ddot{M}_N(S_1, S) = 0 \; \forall S \notin \mathbf{A}$, the LHS of (3.3) reduces to $\sum_{S \in \mathbf{A_j}} \ddot{M}_N(S_1, S)$.

Similarly, for the right hand side (RHS) of (3.3), the super state $S_2 = (v_0, v_1, \dots, v_{N-1})$, $0 \leq v_i \leq N_{max} - 1$, can transit to any super state in set $\mathbf{B} = \{(v'_0, v'_1, \dots, v'_{N-1})\}$, where state $v'_i \in \{v_i - 1, v_i, v_i + 1\}$, i.e., $|\mathbf{B}| \leq 3^{N_{max}}$. Let subsets $\mathbf{B_i} = \mathbf{B} \cap \Omega_i$ and $\mathbf{B_j} = \mathbf{B} \cap \Omega_j$. Since $\ddot{M}_N(S_2, S) = 0 \; \forall S \notin \mathbf{B}$, the RHS of (3.3) reduces to $\sum_{S \in \mathbf{B_j}} \ddot{M}_N(S_2, S)$.

Consider two sequences, S_i and S_j, that are permutations of each other, and define $\varrho(S_i, O_{ij}) = S_j$ to be the permutation operator on sequence S_i under index order O_{ij} that gives S_j, i.e., $S_j = (S_i(O_{ij}(k)))_{k=1}^{|S_j|}$. For example, if $S_i = (1, 0, 2)$ and $S_j = (0, 2, 1)$, then $O_{ij} = (2, 3, 1)$.

Let $S'_1 = (u'_0, u'_1, \dots, u'_{N-1})$ be a super state in subset $\mathbf{A_j}$. Therefore, $\ddot{M}_N(S_1, S'_1) = \prod_{n=0}^{N-1} M(u_i, u'_i)$. Since $S_1, S_2 \in \Omega_i$, there exists an index order O_{12}, s.t. $\varrho(S_1, O_{12}) = S_2$. Additionally, $\exists S'_2 = (v'_0, v'_1, \dots, v'_{N-1})$ s.t. $S'_2 = \varrho(S'_1, O_{12})$. Note that $S'_2 \in \mathbf{B_2}$. As a result, $\ddot{M}_N(S_2, S'_2) = \prod_{n=0}^{N-1} M(v_i, v'_i) = \prod_{n=0}^{N-1} M(u_{O_{12}(n)}, u'_{O_{12}(n)})$. Since the product operation is commutative, then

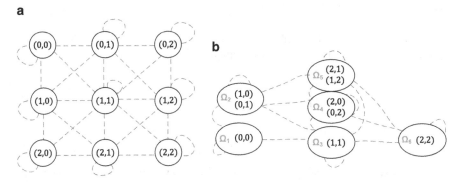

Fig. 3.7 Illustration of (**a**) an N-dimensional Markov chain and (**b**) a lumped Markov chain, both representing the system $\mathbb{X}_N$ for $N = 2, N_{\max} = 2$. The *dashed lines* between two super/lumped states represents a non-zero two-way transition probability in a single time step between the linked states

$\ddot{M}_N(S_2, S_2') = \ddot{M}_N(S_1, S_1')$. In general, $\forall S_1, S_2 \in \Omega_i$ $s.t.\varrho(S_1, O_{12}) = S_2$ and $\forall S_1' \in \mathbf{A_j}, \exists S_2' \in \mathbf{B_j}$ $s.t.S_2' = \varrho(S_1', O_{12})$ and $\ddot{M}_N(S_2, S_2') = \ddot{M}_N(S_1, S_1')$. Hence, $\sum_{S \in \mathbf{A_j}} \ddot{M}_N(S_1, S) = \sum_{S \in \mathbf{B_j}} \ddot{M}_N(S_2, S)$, which ends the proof ∎.

Since a lumped state, $\Omega_i = \{(s_0, s_1, \ldots, s_{N-1})\}, 0 \leq i \leq N_L - 1$, contains all super states that are permutations of the same set of states, the lumped state can be written as a set of those states $\Omega_i = \{s_0, s_1, \ldots, s_{N-1}\}$. Figure 3.7 illustrates how Theorem 3.1 is utilized to reduce the state space of an N-dimensional Markov chain representing the system $\mathbb{X}_N$. The lumped Markov chain in Fig. 3.7b is derived by applying Theorem 3.1 on the N-dimensional Markov chain shown in Fig. 3.7a. Since the N-dimensional Markov chain is irreducible, the lumped Markov chain is also irreducible [13]. The stationary distribution of the lumped Markov chain can be derived from the stationary distribution of the 1D-Markov chain according to the following Corollary [11, 12].

3.2.2 Corollary 3.1: Stationary Distribution of Lumped Group Mobility Model

Corollary 3.1. *Consider a system of N independent copies of a finite, discrete-time, birth-death, irreducible Markov chain, X, with stationary distribution $(\pi_i)_{i=0}^{N_{max}-1}$. The stationary distribution of the lumped Markov chain of Theorem 3.1, representing the system, $\mathbb{X}_N = (X_i)_{i=0}^{N-1}$, follows a multi-nomial distribution with parameters $(\pi_i)_{i=0}^{N_{max}-1}$.*

3.2.2.1 Proof of Corollary 3.1 [11, 12]

Consider the tri-diagonal probability transition matrix of the Markov chain, X, as described in Sect. 3.1. The stationary distribution of the chain, X, is given by

$$\pi_i = \prod_{k=0}^{i-1} \left(\frac{p_k}{q_{k+1}} \right) \pi_0, \quad 1 \le i \le N_{max} - 1 \tag{3.4}$$

where $\pi_0 = \left[1 + \sum_{i=1}^{N_{max}-1} \prod_{k=0}^{i-1} \left(\frac{p_k}{q_{k+1}} \right) \right]^{-1}$. Consider the ith lumped state $\Omega_i = \{s_0, s_1, \ldots, s_{N-1}\}$. Let $N_{D,i}$ be the number of distinct states in lumped state Ω_i, and let $(u_1, u_2, \ldots u_{N_{D,i}})$ and $(n_{u_1}, n_{u_2}, \ldots, n_{N_{D,i}})$ be the sequences of distinct states and their corresponding frequencies, respectively, where $0 \le u_i \le N_{max} - 1$ and $\sum_{i=1}^{N_{D,i}} n_{u_i} = N$. Note that the size of the lumped state is equal to the number of super states that are permutations of each other, i.e., $1 \le |\Omega_i| \le N!, 0 \le i \le N_L - 1$. Therefore, the lumped states result from all possible outcomes of choosing N states from N_{max} different states independently, where choosing state s_i has the probability $\pi_i, 0 \le s_i \le N_{max} - 1$. This is a generalization of the Bernoulli trial problem. Hence, the stationary distribution for the lumped state Ω_i is given by

$$\mho_i = \frac{N!}{\prod_{k=1}^{N_{D,i}} n_{u_k}!} \prod_{k=1}^{N_{D,i}} \pi_{u_k}^{n_{u_k}}. \tag{3.5}$$

That is, the stationary distribution of the lumped Markov chain is multi-nomial, which ends the proof ∎.

At any time instant, the total distance between two non-consecutive vehicles separated by $N - 1$ nodes, is equal to $\sum_{i=0}^{N-1} X_i(m), \forall m \ge 0, N > 1$. Therefore, according to Theorem 3.1, the time variation of the total distance between the two reference vehicles can be described by a lumped Markov chain with lumped states $\Omega_0, \Omega_1, \ldots, \Omega_{N_L-1}$ which represents the system, $\mathbb{X}_N = (X_i)_{i=0}^{N-1}$. In order to characterize the spatiotemporal variations of the VANET topology, the occurrence time of certain events in the distance between two vehicles need to be analyzed. For example, given two nodes separated by $N - 1, N > 1$ nodes that are one-hop away at the 0th time step, what is the time until the distance between the two nodes becomes larger than the communication range, i.e., the occurrence time of event $\sum_{i=0}^{N-1} X_i(m) > R$. To probabilistically characterize such events, transient analysis of the system $\mathbb{X}_N$ should be performed. Divide the lumped states in the partition $\Omega = \{\Omega_0, \Omega_1, \ldots, \Omega_{N_L}\}$ into two sets, initial set Ω_I and event set Ω_L. The event set is defined based on an event of interest to system $\mathbb{X}_N$, for example, the event that the sum of the N distance headways is larger/smaller/equal to a deterministic threshold L. Let T_L be the time interval that the system spends in the initial set Ω_I. Then, T_L is the first occurrence time of system $\mathbb{X}_N$ to hit a lumped state in the event set given that it is initially in set Ω_I. Convert set Ω_L to become absorbing, and let

$\tilde{M}_y$ denote the absorbing probability transition matrix. The probability distribution of T_L is equal to the distribution of the absorbing time and can be calculated by first passage time analysis. Details on the exact calculation of T_L for different thresholds L are given in [11] and are described in Chap. 4.

The size of the state space of the lumped Markov chain can still be large with an increased number of nodes between the two reference vehicles, since $N_L = \frac{(N_{max}+N-1)!}{N!(N_{max}-1)!}$. However, the state space of the absorbing lumped Markov chain, needed to compute occurrence time of the event that the sum of distance headways in $\mathbb{X}_N$ is greater than a certain threshold, is bounded according to the following Corollary [11, 12].

3.2.3 Corollary 3.2: Bounded Absorbing Lumped State Space

Corollary 3.2. *Consider a system of N independent copies of an irreducible Markov chain according in Theorem 3.1, and let the event of interest be that the sum of the states of the N chains be larger than a deterministic threshold N_{th}. The absorbing lumped Markov chain, required to obtain the first occurrence time of the event of interest, has a state space that is bounded by a deterministic function of N_{th}, when $N > N_{th}$.*

3.2.3.1 Proof of Corollary 3.2 [11, 12]

Let the lumped state $\Omega_j = \{s_0, s_1, \ldots, s_{N-1}\}$ be a lumped state such that, if the system enters this state, the event of interest occurs. Then, $\{s_0, s_1, \ldots, s_N\}$ is an N-restricted integer partition of an integer that is greater than or equal to N_{th}. In combinatorics, an integer partition of a positive integer n is a set of positive integers whose sum equals n. Each member of the set is called a *part*. An N-restricted integer partition of an integer n is an integer partition of n into exactly N parts. Therefore, $\forall \Omega_j = \{s_0, s_1, \ldots, s_{N-1}\} \in \Omega_I, \{s_0, s_1, \ldots, s_{N-1}\}$ is an integer partition of an integer that is less than N_{th}. Since, an integer N_{th} can be partitioned into at most N_{th} parts (i.e. when all the parts equal to one) and the order of the N states in the lumped state is not important, the number of lumped states $\in \Omega_I$ when $N > N_{th}$ is equal to that when $N = N_{th}$. Notice that Corollary 3.2 applies only on the lumped Markov chain and not the original N-dimensional one. This ends the proof ∎.

3.2.4 Group Mobility Model Scalability

Consider the scalability of analyzing a system of N distance headways, $\mathbb{X}_N$, to an increased number of distance headways, N. Using the lumped Markov chain, the scalability of analyzing system $\mathbb{X}_N$ is improved for:

- **the steady-state analysis**—The problem of finding the stationary distribution of a system of distance headway is of constant computational complexity, i.e., $O(1)$, where $O(.)$ is the big O notation, with respect to N (according to Corollary 3.1); and

- **the transient analysis**—The computational complexity of the transient analysis (i.e, the first passage time analysis) is dependent on the state space size of the considered Markov chain. According to Corollary 3.2, the state space size of the absorbing lumped Markov chain is upper bounded by the total number of integer partitions of all integer that are less than N_{th} as discussed in Sect. 3.2.3.1.

Consider the problem of finding the probability distribution of the first passage time for system $\mathbb{X}_N$ to hit a state such that the sum of the state indices of $\mathbb{X}_N$ is greater than or equal to threshold N_{th}. The computational complexity of this problem is $O(\mathfrak{D}(|\mathbb{S}_N|)\log(m))$ where m is the time step value at which one wishes to stop the probability distribution calculation, and $\mathfrak{D}(|\mathbb{S}_N|)$ is the computational complexity of the matrix multiplication algorithm for a matrix of size $|\mathbb{S}_N| \times |\mathbb{S}_N|$. A naive multiplication algorithm has $\mathfrak{D}(|\mathbb{S}_N|) = O(|\mathbb{S}_N|^3)$, where $|\mathbb{S}_N|$ is the state space size of the Markov chain representing the system. The value of $|\mathbb{S}_N|$ equals to $N^{N_{\max}}, N_L$, and $\tilde{N}_L$ when using the N-dimensional, lumped, and absorbing lumped Markov chains, respectively. If $N_{\text{th}} \leq N_{\max}$, then

$$\tilde{N}_L = 1 + \sum_{i=0}^{N_{\text{th}}-1} I_N(i) \tag{3.6}$$

where $I_k(n)$ is the number of integer partitions of n into at most k parts and is given by $I_k(n) = I_k(n-k) + I_{k-1}(n)$, where $I_0(n) = 0$, $I_1(n) = 1, I_2(n) = 1 + \lfloor \frac{n}{2} \rfloor$, and $I_3(n) = 1 + \lfloor \frac{n^2+6n}{12} \rfloor$.

Figure 3.8 shows the state space reduction using the proposed lumped Markov chain.

3.3 Summary

This chapter presents a new stochastic microscopic mobility model that describes the distance headway of individual vehicles. The model captures the time variations in a distance headway based on a discrete-time Markov chain that preserves the realistic dependency of distance headway changes at consecutive time steps. This

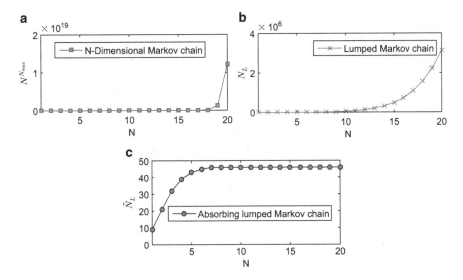

Fig. 3.8 The state space size of a Markov chain representing a system of N Markov chains (distance headways), $\mathbb{X}_N$, with $N_{\max} = 9$ when the system $\mathbb{X}_N$ is represented by (**a**) an N-dimensional Markov chain, (**b**) a lumped Markov chain according to Theorem 3.1, and (**c**) an absorbing lumped Markov chain according to Corollary 3.2 with $N_{th} = 8$

dependency increases with the vehicle density, which is consistent with highway data patterns from empirical NGSIM and simulated VISSIM data sets. Using lumpability theory, the proposed mobility model is extended to describe the behavior of a group of vehicles on road with reduced dimensionality. Computational complexity analysis of the proposed group mobility model shows its scalability to a large number of vehicles. The proposed mobility model is a promising candidate vehicle mobility model to be utilized for traceable mathematical analysis in VANETs.

References

1. X. Chen, L. Li, and Y. Zhang, "A markov model for headway/spacing distribution of road traffic," *IEEE Trans. Intelligent Transportation Systems*, vol. 11, no. 4, pp. 773–785, 2010.
2. Q. Gong, S. Midlam-Mohler, V. Marano, and G. Rizzoni, "An iterative markov chain approach for generating vehicle driving cycles," *SAE Inter. J. Engines*, vol. 4, no. 1, pp. 1035–1045, 2011.
3. K. Abboud and W. Zhuang, "Analysis of communication link lifetime using stochastic microscopic vehicular mobility model," in *Proc. IEEE Globecom*, 2013, pp. 383–388.
4. K. Abboud and W. Zhuang, "Stochastic analysis of single-hop communication link in vehicular ad hoc networks," *IEEE Trans. Intelligent Transportation Systems*, vol. 15, no. 5, pp. 2297–2307, 2014.
5. A. May, *Traffic Flow Fundamentals*. Prentice Hall, 1990.

6. M. Krbalek and K. Kittanova, "Theoretical predictions for vehicular headways and their clusters," *Physics: Data Analysis, Statistics and Probability (arXiv)*, 2012.

7. L. Li, W. Fa, J. Rui, H. Jian-Ming, and J. Yan, "A new car-following model yielding log-normal type headways distributions," *Chinese Physics B*, vol. 19, no. 2, 2010.

8. Next generation simulation community, vehicle trajectory data sets. http://ngsim-community. org/. Accessed Nov. 4, 2012.

9. M. Fellendorf and P. Vortisch, "Microscopic traffic flow simulator VISSIM," *Fundamentals of Traffic Simulation*, pp. 63–93, 2010.

10. PTV, "VISSIM 5.40 user manual," *Karlsruhe, Germany*, 2012.

11. K. Abboud and W. Zhuang, "Stochastic modeling of single-hop cluster stability in vehicular ad hoc networks," *IEEE Trans. Vehicular Technology*, 2015 (to appear).

12. K. Abboud and W. Zhuang, "Impact of node mobility on single-hop cluster overlap in vehicular ad hoc networks," in *Proc. ACM MSWiM*, 2014, pp. 65–72.

13. P. Buchholz, "Exact and ordinary lumpability in finite Markov chains," *J. Applied probability*, vol. 31, no. 1, pp. 59–75, 1994.

Chapter 4
Spatiotemporal Network Topology Analysis

Network topology in VANETs is subject to fragmentations and frequent changes due to vehicle mobility. Moreover, VANETs are susceptible to vehicle density variations from time to time throughout the day. This imposes new challenges in maintaining a connection between vehicular nodes. As discussed in Sect. 1.3, the frequent changes in VANET topology may degrade the performance of network protocols. In this chapter, the spatiotemporal variations in VANET topology are analyzed. Two parameters are used to describe the network topology, the communication link and vehicle's neighbors. Firstly, the length of the communication link is analyzed using mesoscopic vehicle mobility models. Secondly, the proposed microscopic mobility model is utilized to derive the communication link duration. Thirdly, the lumped Markov chain is relaxed to an edge-lumped Markov chain and the probability distribution of the time period between successive changes in vehicle's one-hop neighborhood is derived. Furthermore, queueing theory is utilized to model the limiting behavior of the common VNs of two reference vehicles that are two-hop away. The overlapping region of the coverage ranges of the two-hop vehicles is modeled as a storage buffer in a two-state random environment. Using G/G/1 queuing theory, the steady-state distribution of the number of common VNs is approximated. Numerical results are presented to evaluate the proposed models, which demonstrate a close agreement between analytical and simulation results.

4.1 Communication Link Lifetime

In this section, the probability distribution of the communication link length and the communication link duration is derived using mesoscopic and microscopic distance headway models, respectively. Assume the mesoscopic vehicle models described in Sect. 2.2. The hop length (or the link length), denoted by H, is the distance from a

© The Author(s) 2015
K. Abboud, W. Zhuang, *Mobility Modeling for Vehicular Communication Networks*,
SpringerBriefs in Electrical and Computer Engineering,
DOI 10.1007/978-3-319-25507-1_4

reference node to the furthest node within the transmission range of the reference node, which is upper bounded by the transmission range R. Given a mesoscopic model, the distance headways are i.i.d. with probability density function $f_X(x)$ and cdf $F_X(x)$. Let $C(l)$ be the event that there exists at least one node within distance l from a reference node. The event $C(l)$ occurs with probability $F_X(l)$. Let $C^c(l)$ be the complement of event $C(l)$, i.e., the event that there are no nodes within distance l from a reference node. Then, the cdf of H is given by Hou and Li [1]

$$F_H(h) = \frac{P(C^c(R-h), C(h))}{P(C(R))}. \tag{4.1}$$

The pdf can then be calculated by $f_H(h) = \frac{d}{dh_r} F_H(h)$. For a low vehicle density, the distance headway is exponentially distributed with pdf given in (2.1). The pdf of the corresponding hop length is given by Cheng and Robertazzi [2]

$$f_H(h) = \frac{e^{-\frac{(R-h)}{\mu}}}{\mu(1 - e^{-\frac{R}{\mu}})}, \quad 0 < h < R \tag{4.2}$$

which is a scaled exponential distribution truncated at R. For an intermediate vehicle density, the distance headways are i.i.d., each following the Pearson type III pdf in (2.3). The cdf for the first hop length can be derived from (4.1) and the corresponding pdf is found to be [3]

$$f_H(h) = \frac{f_X(R-h)\gamma(z, \lambda(h-\alpha)) + f_X(h)\Gamma(z, \lambda(R-h-\alpha))}{\gamma(z, \lambda(R-\alpha))}, \quad \alpha \leq h < R - \alpha \tag{4.3}$$

where $\gamma(z, x) = \int_0^x t^{z-1} e^{-t} dt$ and $\Gamma(z, x) = \int_x^\infty t^{z-1} e^{-t} dt$ are the lower and the upper incomplete gamma functions, respectively, and $f_X(\cdot)$ is given by (2.3). The derivation for $f_H(h)$ is given in the Appendix of [3].

For a high vehicle density, the distance headways are i.i.d., each following the Gaussian pdf in (2.2). Using the cdf of the Gaussian distribution, $F_X(x) = \frac{1}{2}\left(1 + \mathrm{erf}\left(\frac{x-\mu}{\sqrt{2}\sigma}\right)\right)$, the cdf for the hop length can be derived from (4.1) and is given by Abboud and Zhuang [3]

$$f_H(h) = \frac{1}{\sqrt{2\pi}\sigma}\left(1 + \mathrm{erf}\left(\frac{R-\mu}{\sqrt{2}\sigma}\right)\right)^{-1} \times \left[e^{-\frac{(h-\mu)^2}{2\sigma^2}}\right.$$
$$\times \left(1 - \mathrm{erf}\left(\frac{R-h-\mu}{\sqrt{2}\sigma}\right)\right) + e^{-\frac{(R-h-\mu)^2}{2\sigma^2}}$$
$$\left. \times \left(1 + \mathrm{erf}\left(\frac{h-\mu}{\sqrt{2}\sigma}\right)\right)\right], \quad 0 < h < R \tag{4.4}$$

where $\mathrm{erf}(\cdot)$ is the error function, given by $\mathrm{erf}(x) = \frac{2}{\sqrt{\pi}} \int_0^x e^{-t^2} dt$.

Consider a communication hop from an arbitrary reference node in the direction of the vehicle traffic flow. The reference node is one hop away from all the nodes within a distance less than R (assuming that an on/off link depends only on the distance between the nodes). Define the communication link lifetime between two nodes as the first time step at which the distance between the two nodes is larger than or equal to R, given that the distance between them is less than R at the 0th time step. For any node within R from the reference node, the communication link lifetime is at least equal to the that of the furthest node from the reference vehicle (referred to as *hop edge node*). In order to analyze the communication link lifetime, the microscopic mobility model proposed in Chap. 3 is utilized. A study of the communication link lifetime between a reference vehicle and its hop edge node (i.e., link-edge nodes) is presented in the following.

4.1.1 Consecutive Link-Edge Nodes

Consider a reference vehicle that has one vehicle in its communication range. That is, the reference vehicle and its hop edge node are consecutive. Therefore, the time variations in the communication link between the reference vehicle and its hop edge node is characterized by the one dimensional discrete-time Markov chain, which describes the distance headway between the consecutive vehicles and proposed in Sect. 3.1. Let the distance headway be in state $s_j, 0 \leq j < N_R$ at the 0th time step. Then the communication link lifetime is the first passage time for the distance headway to reach state s'_j s.t. $N_R \leq j' \leq N_{max}$. In the following, the general problem of calculating the probability distribution of the first passage time between any two states s_j and s'_j, $0 \leq j, j' \leq N_{max} - 1$ is considered.

Let $T_{j,j'}, 0 \leq j, j' \leq N_{max} - 1$, be the first passage time of the distance headway X to state j' given that the distance headway is in state j at the 0th time step, i.e., $T_{j,j'} = \min\{m > 0; X_i(m) = x_{j'}, X_i(0) = x_j\}, 0 \leq j \leq N_{max} - 1$. Let M' be an $N_{max} \times N_{max}$ matrix equal to M with $q_{N_{max}-1} = 0$ and $r_{N_{max}-1} = 1$. Let $\{\lambda_u\}_{u=0}^{N_{max}-2}$ be the $N_{max}-1$ non-unit eigenvalues of M'. The first passage time to state $N_{max}-1$, given that $X_i(0) = x_0$, is the sum of $N_{max} - 1$ independent geometric random variables, each with a mean equal to $\frac{1}{1-\lambda_u}$ [4]. The probability generating function of $T_{0,N_{max}-1}$ is given by

$$Q_{T_{0,N_{max}-1}}(v) = \prod_{u=0}^{N_{max}-2} \left[\frac{(1-\lambda_u)v}{1-\lambda_u v} \right]. \tag{4.5}$$

The pmf of $T_{0,N_{max}-1}$ is then calculated by $P_{T_{0,N_{max}-1}}(m) = \frac{Q_{T_{0,N_{max}-1}}^{(m)}(0)}{m!}$, where $Q_{T_{0,N_{max}-1}}^{(m)}(0)$ is the value of the mth derivative of $Q_{T_{0,N_{max}-1}}(v)$ at $v = 0$.

Let $M^{(j)}$ be a $(j + 1) \times (j + 1)$ matrix, $0 < j < N_{max} - 1$, equal to the upper left $(j + 1) \times (j + 1)$ portion of matrix M with $q_j = 0$ and $r_j = 1$. The first passage

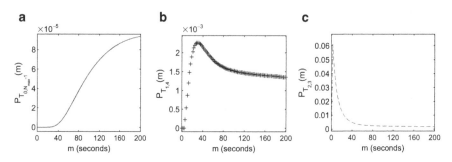

Fig. 4.1 Probability mass function of the first passage time for (**a**) $T_{1,N_{max}}$, (**b**) $T_{1,4}$, and (**c**) $T_{2,3}$, with mean values of 19.8×10^3, 1.24×10^3, and 318.4 s, respectively, with parameters $N_{max} = 9$, $L_s = 20$ m, $\tau = 2$ s, $X_{max} = 160$ m, $\beta = 0.66$, $p = 0.12$, and $q = 0.26$

time of the distance headway to state j, given that the initial distance headway is in state 0, has a probability generating function $Q_{T_{0,j}}(v) = \prod_{u=0}^{j-1} \left[\frac{(1-\lambda_u^{(j)})v}{1-\lambda_u^{(j)}v} \right]$, where $\lambda_u^{(j)}$, $u = 0, 1, \ldots, j-1$, are the j non-unit eigenvalues of $M^{(j)}$. Note that the distance headway cannot move to state $j'(> j)$ before passing through state j in a birth and death process. Using $T_{0,j'} = T_{0,j} + T_{j,j'}$, the passage time to state j' given that the initial distance headway is in state j, $0 \leq j < j' \leq N_{max} - 1$, can be calculated. The probability generating function of $T_{j,j'}$ is $Q_{T_{j,j'}}(v) = \frac{E\left[v^{T_{0,j'}}\right]}{E\left[v^{T_{0,j}}\right]}$, and is calculated by

$$
Q_{T_{j,j'}}(v) = v^{j'-j} \frac{\prod_{u=0}^{j'-1} \left[\frac{(1-\lambda_u^{(j')})}{1-\lambda_u^{(j')}v} \right]}{\prod_{u=0}^{j-1} \left[\frac{(1-\lambda_u^{(j)})}{1-\lambda_u^{(j)}v} \right]}.
\tag{4.6}
$$

Figure 4.1 plots the pmf's of $T_{0,N_{max}-1}$, $T_{1,4}$ and $T_{2,3}$ for a nine-state distance headway model. The pmf's are evaluated using MAPLE [5]. Figure 4.1 shows that the probability of the first passage time, $T_{j,j'}$, taking on a small value decreases as the number of states, $|j' - j|$, increases. The communication link lifetime between a reference vehicle and its hop edge node given that $X(0) \in s_j$, $0 \leq j < N_R$ equals $T_{j,j'}$ for $j' = N_R$.

4.1.2   Non-consecutive Link-Edge Nodes

Consider a vehicle with more than one node in its one-hop neighborhood from one side. Let N_H be the number of distance headways between the reference vehicle and its hop edge node at the 0th time step (i.e., the two nodes are separated by $N_H - 1$ nodes, $N_H > 1$). The distance between a reference node and its hop edge node is equal to the sum of the distance headways between the two nodes. Label the nodes with IDs: $\{0, 1, \ldots N_H\}$, where the reference node has ID 0, and the hop edge node has ID N_H. Therefore, $R \leq \sum_{i=0}^{N_H} X_i(0) < R + X_{N_H}(0)$. A node and its hop edge node remain connected until $\sum_{i=0}^{N_H-1} X_i(m) \geq R$ at some time step m which is the communication link lifetime.

The sum of N_H i.i.d. distance headways, where each headway, X_i, $0 \leq i < N_H$, is a birth and death Markov chain as illustrated in Fig. 3.1, is an N_H-dimensional Markov chain. The complexity of this Markov chain is obvious especially when N_H is not small, since a non-zero transition probability to a non-neighboring state is possible. It is possible to tackle the problem of large state space of the N_H-dimensional Markov chain by defining link disconnection events and then apply the transient analysis, as done in [3]. However, using the group mobility model proposed in Sect. 3.2 provides lower computational complexity in calculating the distribution of communication link lifetime as discussed in Sect. 3.2.4.

4.1.2.1 Time to the First Link Breakage

Consider the communication link between a reference vehicle and its hop edge node separated by $N_H - 1$ nodes. At any time instant, the communication link length is equal to the total distance between the nodes, i.e., $\sum_{i=0}^{N_H-1} X_i(m)$, $\forall m \geq 0$. Therefore, according to Theorem 3.1, the time variation of the communication link between the two nodes can be described by a lumped Markov chain with lumped states $\Omega_0, \Omega_1, \ldots, \Omega_{N_L-1}$ which represents the system, $\mathbb{X}_{N_H} = (X_i)_{i=0}^{N_H-1}$. Furthermore, divide the lumped states into two sets, Ω_I and Ω_R. A lumped state $\Omega_i = \{s_0, s_1, \ldots, s_{N_H-1}\}$ belongs to Ω_I and to Ω_R if $\sum_{i=0}^{N_H-1} s_i < N_R$ and $\sum_{i=0}^{N_H-1} s_i \geq N_R$, respectively. Let the system of the distance headways between the two nodes be initially in super state I_0, i.e., $\mathbb{X}_{N_H}(0) \in I_0$, s.t. $I_0 \in \Omega_k \in \Omega_I$, $0 \leq k \leq N_L - 1$. Let the time period until the communication link breaks be $T_{R0}(\Omega_k)$, given that the distance headways between them are initially in states $I_0 \in \Omega_k$. Then, this time period is equal to the first passage time for the system, $\mathbb{X}_{N_H}$, to transit from the lumped state Ω_k to any lumped state $\Omega_{k'}$, s.t. $\Omega_{k'} \in \Omega_R$. That is, $T_{R0}(\Omega_k) = \min\left\{m > 0; \ \mathbb{X}_{N_H}(m) \in (k_0, k_1, \ldots, k_{N_H-1}), \sum_{i=0}^{N_H-1} k_i \geq N_R \mid \mathbb{X}_{N_H}(0) \in I_0\right\}$. Let M_{N_H} be the transition probability matrix of the lumped Markov chain describing $\mathbb{X}_{N_H}$. One way to find the first passage time is to force the lumped states in Ω_R to become absorbing, i.e., set the probability of returning to the same lump state, Ω_i, within one time step to one $\forall \Omega_i \in \Omega_R$. Furthermore, let all the lumped states in Ω_R be merged into one single absorbing state and let it be the last $(\tilde{N}_L - 1)$th state, where $\tilde{N}_L$ is the number of states in the new absorbing lumped Markov chain. The transition probability matrix of the new absorbing lumped Markov chain, $\tilde{M}_{N_H}$, is derived from M_{N_H} as follows: $\tilde{M}_{N_H}(\Omega_i, \Omega_j) = M_{N_H}(\Omega_i, \Omega_j) \ \forall i, j, \ s.t. \ \Omega_i, \Omega_j \in \Omega_I$, $\tilde{M}_{N_H}(\Omega_i, \Omega_{N_L-1}) = \sum_j M_{N_H}(\Omega_i, \Omega_j) \ \forall i, j, \ s.t. \ \Omega_i \in \Omega_I$ and $\Omega_j \in \Omega_R$. Let $T_{R0}(\Omega_k)$ denote the time interval from the 0th time step till the first time instant that the two vehicles are no longer one-hop neighbors, given that the distance headways are in super state $I_0 \in \Omega_k$. The cdf of $T_{R0}(\Omega_k)$ is given by

$$F_{T_{R0}(\Omega_k)}(m) =$$

$$\tilde{M}_{N_H}(\Omega_k, \Omega_{\tilde{N}_L-1}) + \sum_{\substack{j \\ \Omega_j \in \Omega_I}} \tilde{M}_{N_H}(\Omega_k, \Omega_j) F_{T_{R0}(\Omega_j)}(m-1), \quad m \geq 1 \qquad (4.7)$$

where $F_{T_{R0}(\Omega_k)}(0) = 0$. Equation (4.7) calculates the cdf of $T_{R0}(\Omega_k)$ recursively. Since $F_{T_{R0}(\Omega_k)}(m) = \sum_{n=1}^{m} P_{T_{R0}(\Omega_k)}(m)$, the first term in (4.7) corresponds to the absorption probability within one time step given that the system is initially in lumped state Ω_k, i.e., $F_{T_{R0}(\Omega_k)}(1) = \tilde{M}_{N_H}(\Omega_k, \Omega_{\tilde{N}_L-1})$. The second term in (4.7) corresponds to $\sum_{n=2}^{m} P_{T_{R0}(\Omega_k)}(m)$ which is the absorption probability within $(m-1)$ time steps given that the system transited from Ω_k to $\Omega_j \in \Omega_I$ within one time step.

The size of the state space of the lumped Markov chain can still be large with an increased number of nodes between the reference vehicle and its hop edge node, since $N_L = \frac{(N_{max}+N_H)!}{(N_H+1)!(N_{max}-1)!}$. However, the state space of the absorbing lumped Markov chain, needed to compute the time period until the first link breakage, is bounded according Corollary 3.2.

Up until now, the communication link considered is initially in a specific super state. In reality, the initial state of the link between a vehicle and its hop edge node is a random variable. For a given N_H, since the distance headways are assumed stationary at the 0th time step, the probability that the communication link is initially

in Ω_i is given by $\mho_i \Big/ \left[\sum_{j,\Omega_j \in \Omega_I} \mho_j \right]$ where $\mho_i$ is given by (3.5) in Sect. 3.2.2.1.

Using the law of total probability, the cdf of the time for the first breakage in the communication link between a vehicle and its hop edge node is given by

$$F_{T_{R0}}(m) = \frac{\sum_{\substack{j \\ \Omega_j \in \Omega_I}} \mho_j F_{T_{R0}(\Omega_j)}(m)}{\sum_{\substack{i \\ \Omega_i \in \Omega_I}} \mho_i}, \quad m = 1, 2, \ldots \qquad (4.8)$$

4.1.2.2   Time Period Between Successive Changes in Link-State

A communication link between a vehicle and its hop edge node can fluctuate between two link states: connected and disconnected. A communication link can be in two states, connected or disconnected. The preceding subsection has presented analysis of the time interval during which the communication link between a reference vehicle and its hop edge node remains connected from the initial time step 0th. During this time interval, the communication link state remains unchanged. The communication link-state may change and become disconnected and then may change again and become connected. As a result, the time period between two consecutive changes in link-state equals (1) the connection time period when the link-state changes from connected to disconnected, or (2) the disconnection time period when the link-state changes from disconnected to connected. This subsection focuses on the connection time period due to its importance in the design of network protocols. Note that the connection time period, for a reference vehicle and its hop edge node, equals the time interval calculated from the instant they become connected until the time instant when they are no longer connected, which is also equal to the communication link lifetime. However, this connection time (second communication link duration) may not be equal to T_{R0}, since the initial state may not be the same as that at the 0th time step. This period is referred to as communication link duration, denoted by T_R.

To derive the distribution of T_R, the same approach used to find the distribution of T_{R0} can be used. Notice that the absorbing lumped Markov chain is the same as that used to calculate the distribution of T_{R0}. The only difference is the distribution of the initial state, I_0. One way to find the distribution of I_0 at the time when the second connected link-state occurs is as follows:

- Make the lumped states in set Ω_R absorbing, without combining them into one absorbing state. The corresponding transition probability matrix, M''_{N_H}, is equal to M_{N_H} with $M''_{N_H}(\Omega_j, \Omega_i) = 0$ and $M''_{N_H}(\Omega_j, \Omega_j) = 1 \forall i, j,\ s.t.\ \Omega_j \in \Omega_R$;
- Calculate the absorbing probability δ_j for each absorbing lumped state $\Omega_j \in \Omega_R$ by

$$\delta_j = \sum_{\substack{i \\ \Omega_i \in \Omega_I}} \frac{\mho_i}{\sum_{\substack{k \\ \Omega_k \in \Omega_I}} \mho_k} \lim_{m \to \infty} M''^{(m)}_{N_H}(\Omega_i, \Omega_j) \tag{4.9}$$

where $M''^{(m)}_{N_H}(\Omega_i, \Omega_j)$ denotes the (Ω_i, Ω_j)th entry of the mth power of matrix M''_{N_H};
- Form another absorbing Markov chain by making the lumped states in set Ω_I absorbing, without combining them into one absorbing state. The corresponding transition probability matrix, M'_{N_H}, is equal to M_{N_H} with $M'_{N_H}(\Omega_i, \Omega_j) = 0$ and $M'_{N_H}(\Omega_i, \Omega_i) = 1 \forall i, j,\ s.t.\ \Omega_i \in \Omega_I$;
- Calculate the absorbing probability ψ_i for each absorbing lumped state $\Omega_i \in \Omega_I$ by

$$\psi_i = \sum_{\substack{j \\ \Omega_j \in \Omega_R}} \delta_j \lim_{m \to \infty} M'^{(m)}_{N_H}(\Omega_j, \Omega_i). \tag{4.10}$$

The probability that the distance headways between the reference vehicle and its hop edge node are in state $\Omega_i \in \Omega_I$ at the time when the second connected link-state occurs is equal to ψ_i. Therefore, the cdf of the communication link lifetime is given by

$$F_{T_R}(m) = \sum_{\substack{i \\ \Omega_i \in \Omega_I}} \psi_i F_{T_{R0}(\Omega_i)}(m), m = 1, 2, \dots \tag{4.11}$$

where $F_{T_{R0}(\Omega_i)}(m)$ is given by (4.7). However, using this approach, one loses the advantage of having a single absorbing state and, therefore, a bounded state space (according to Corollary 3.2). Therefore, an approximation is proposed for the distribution of the system initial state at the time when the second connected link-state occurs, ψ_i, as follows

$$\psi_i \approx \frac{\mho_i \tilde{M}_{N_H}(\Omega_i, \Omega_{\tilde{N}_L-1})}{\sum\limits_{\substack{i \\ \Omega_i \in \Omega_I}} \mho_i \tilde{M}_{N_H}(\Omega_i, \Omega_{\tilde{N}_L-1})}. \tag{4.12}$$

The approximated ψ_i for lumped state $\Omega_i (\in \Omega_I)$ is equal to its stationary probability weighted with the absorption probability within one time step. Notice that this weight eliminates all the lumped states $\Omega_i \in \Omega_I$ that are not directly accessible from states in Ω_R. Figure 4.2 illustrates an example for a lumped Markov chain, where the directly accessible lumped states are those connected by solid lines, i.e. Ω_{I1} and Ω_{R1}. When the link-state changes from disconnected to connected, the only possible states to be reached first are those in Ω_{I1}.

The average connection-time period (communication link lifetime) is given by Rubino and Sericola [6]

$$E[T_R] = \Psi\left(I - \tilde{M}_{N_H}\right)^{-1} M_1 \tag{4.13}$$

where Ψ is a row vector of size $\tilde{N}_L$ in which the jth element equals ψ_j, I is the identity matrix of size equal to that of $\tilde{N}_L$, and M_1 is a column vector of ones with

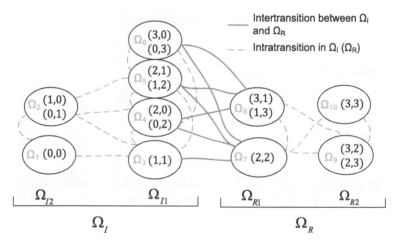

Fig. 4.2 Illustration of a lumped Markov chain representing the system $\mathbb{X}_{N_H}$ for $N_H = 2, N_R = 4, N_{max} = 3$. A line between two lumped states represents a non-zero two-way transition probability in a single time step between the linked states. There exist non-zero transition probabilities between subsets of Ω_{I1} and Ω_{R1}

size $\tilde{N}_L$. The second moment of the communication link lifetime is given by Rubino and Sericola [6]

$$E[T_R^2] = 2\Psi\tilde{M}_{N_H}\left(I - \tilde{M}_{N_H}\right)^{-2}M_1 + E[T_R]. \tag{4.14}$$

4.2   Temporal Variations in Vehicle's One-Hop Neighborhood

The switching of communication links between connection and disconnection is associated with vehicles moving in and out of the communication range of other vehicles, resulting in changes in vehicles one-hop neighborhood. This section investigates the microscopic vehicle mobility impact on vehicles one-hop neighborhood. Due to relative vehicle mobility, two events result in changes to the vehicle's one-hop neighborhood: (1) a vehicle leaving the reference vehicle's coverage range, and (2) a vehicle entering the reference vehicle's coverage range and becoming a VN. Let e_{o_r} and e_{o_l} denote the events that a vehicle leaves the coverage range from the right side and the left side of the reference vehicle, respectively. Let e_{i_r} and e_{i_l} denote the events that a vehicle enters the neighborhood from the right side and the left side of the reference vehicle, respectively. Figure 4.3 illustrates these events. Consider the time for the first change in vehicle-neighborhood to occur from the 0th time step, and denote this time by T_{VN0}. This time is equivalent to the first occurrence time of one of the four events, i.e., $T_{VN0} = T(e_{o_r} \cup e_{i_r} \cup e_{o_l} \cup e_{i_l})$, where $T(e)$ denotes the first occurrence time of event e. Furthermore, let $T_{VN0_r} = T(e_{o_r} \cup e_{i_r})$ and $T_{VN0_l} = T(e_{o_l} \cup e_{i_l})$ be the first occurrence time of the first change in vehicle-neighborhood due to a vehicle leaving and entering the coverage range from the right and the left side of the reference vehicle, respectively. Therefore, $T_{VN0} = \min\{T_{VN0_r}, T_{VN0_l}\}$. Since T_{VN0_r} and T_{VN0_l} are independent, the cdf of the time for the first change in vehicle-neighborhood to occur after initial time (0th) is given by $F_{T_{VN0}}(m) = 1 - (1 - F_{T_{VN0_r}}(m))(1 - F_{T_{VN0_l}}(m))$. Notice that T_{VN0_r} and T_{VN0_l} are i.i.d.. Therefore, the calculation of only one of them, say T_{VN0_r}, is presented in the following.

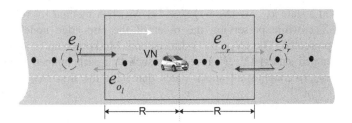

Fig. 4.3 Illustration of the events that cause changes in vehicle's one-hop neighborhood

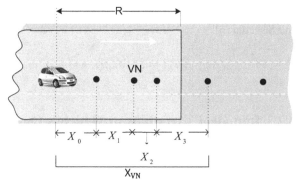

Fig. 4.4 A vehicle's one-hop neighborhood with $N_{VN} = 3$ and $\mathbb{X}_{VN} = (X_0, X_1, X_2, X_3)$

4.2.1 Time to the First Change in Vehicle'S One-Hop Neighborhood

Let N_{VN} be the number of VNs on one side of the reference vehicle, and assume that $N_{VN} > 0$.[1] Let $\mathbb{X}_{VN} = (X_i)_{i=0}^{N_{VN}}$ be the sequence of distance headways of the reference vehicle and the N_{VN} nodes as illustrated in Fig. 4.4, where $\mathbb{X}_{VN}(m) \subseteq (s_0, s_1, \ldots, s_{N_{VN}}) \equiv [X_i(m) \in s_i, \forall i \in [0, N_{VN}]]$. The system, $\mathbb{X}_{VN}$, can be represented by an $(N_{VN} + 1)$-dimensional Markov chain. Suppose that set $\mathbb{X}_{VN}$ is in super state $I_{VN} = (k_0, k_1, \ldots, k_{N_{VN}})$ at the 0th time step, $s.t.$, $\sum_{i=0}^{N_{VN}-1} k_i < N_R$, $\sum_{i=0}^{N_{VN}} k_i \geq N_R$, and $I_{VN} \in \Omega_k$. Let the time period until a node enters/leaves the coverage range from one side of the reference vehicle be $T_{VN0_r(\Omega_k)}$, given that $\mathbb{X}_{VN} \in I_{VN} \in \Omega_k$. Then this time period is equal to the first passage time for the system, $\mathbb{X}_{VN}$, to transit from super state I_{VN} to a super state $(k_0', k_1', \ldots, k_{N_{VN}}')$ such that $\sum_{i=0}^{N_{VN}} k_i' < N_R$ (i.e., a node enters the coverage range) or $\sum_{i=0}^{N_{VN}-1} k_i' \geq N_R$ (i.e., a node leaves the coverage range). That is, $T_{VN0_r(\Omega_k)} = \min \Big\{ m > 0; \mathbb{X}_{VN}(m) \in (k_0',$
$k_1', \ldots, k_{N_{VN}}'), \big\{ \sum_{i=0}^{N_{VN}} k_i' < N_R \cup \sum_{i=0}^{N_{VN}-1} s_i \geq N_R \big\} \mid \mathbb{X}_{VN} \in I_{VN} \Big\}$.

Since the change in vehicle's one-hop neighborhood occurs at the edge of the coverage range of the reference vehicle, the value of $X_{N_{VN}}$ in the system, $\mathbb{X}_{VN}$, is critical to identify the change. Notice that, initially, the distance headway $X_{N_{VN}}$ can only be in a state $k_{N_{VN}} \in [N_R - \sum_{i=0}^{N_{VN}-1} k_i, N_{\max}]$. A method for representing the system $\mathbb{X}_{VN}$ while preserving the order of the hop edge node's distance headway is presented. The $(N_{VN} + 1)$-dimensional Markov chain is lumped into partitions (lumped states) $\Omega_0', \Omega_2', \ldots \Omega_{N_L-1}'$, such that each lumped state $\Omega_i' = \{(s_0, s_1, \ldots, s_{N_{VN}})\}$ contains all super states that have the first N_{VN} states, i.e.,

---

[1]When $N_{VN} = 0$, the problem reduces to a single distance headway, with only the event of a node entering the coverage range causing change in vehicle's one-hop neighborhood. In this case, the first passage time analysis for one dimensional chain can be used.

$(s_0, s_1, \ldots, s_{N_{VN}-1})$, as permutations of each other.[2] This chain is referred to as *edge lumped Markov chain*. Furthermore, divide the lumped states into three sets, Ω_I, Ω_O and Ω_E, such that a lumped state $\Omega'_i = \{(s_0, s_2, \ldots, s_{N_{VN}})\}$ belongs to (1) Ω_I, if $\sum_{i=0}^{N_{VN}-1} s_i < N_R$, and $\sum_{i=0}^{N_{VN}} s_i \geq N_R$; (2) Ω_L, if $\sum_{i=0}^{N_{VN}-1} s_i \geq N_R$; and (3) Ω_E, if $\sum_{i=0}^{N_{VN}} s_i < N_R$. Let $M_{N_{VN}}$ be the transition probability matrix of the described lumped Markov chain. The occurrence time for the first change in vehicle's one-hop neighborhood, $T_{VN1_r(\Omega_k)}$, is the first passage time for system $\mathbb{X}_{VN}$ to transit from super state $I_{VN} \in \Omega_k \in \Omega_I$ to any state in Ω_O (i.e., when a node leaves the coverage range) or Ω_E (i.e., when a node enters the coverage range). To find the distribution of $T_{VN0_r(\Omega_k)}$, the lumped states in Ω_E and Ω_O are forced to become one absorbing state. Following the same steps as in Sect. 4.1.2.1, the cdf of $T_{VN0_r(\Omega_k)}$ can be derived as

$$F_{T_{VN0_r(\Omega_k)}}(m) = \tilde{M}_{N_{VN}}(\Omega_k, \Omega_{\tilde{N}_L-1}) +$$
$$\sum_{\substack{j \\ \Omega_j \in \Omega_I}} \tilde{M}_{N_{VN}}(\Omega_k, \Omega_j) F_{T_{VN0_r(\Omega_j)}}(m-1), \quad m \geq 1 \quad (4.15)$$

where $\tilde{M}_{N_{VN}}$ is the probability transition matrix of the new absorbing lumped Markov chain with $\tilde{N}_L$ states, such that the $(\tilde{N}_L - 1)$th state is the single absorbing state containing all states in Ω_E and Ω_O.

For a random initial state of $\mathbb{X}_{VN}$, the probability that $\mathbb{X}_{VN}$ is initially in lumped state $\Omega'_i = \{(s_0, s_2, \ldots, s_{N_{VN}})\}$ is given by $\frac{\mho_i}{\sum_{j|\Omega_j \in \Omega_I} \mho_j} \times \frac{\pi_{s_{N_{VN}}}}{\sum_{k=K_i}^{N_{max}} \pi_k}$, $K_i = N_R - \sum_{u=0}^{N_{VN}-1} s_u$, where $\mho_i$ is the stationary distribution of lumped state $\Omega_i = \{(s_0, s_2, \ldots, s_{N_{VN}-1})\}$ of the N_{VN}-dimensional Markov chain lumped according to Theorem 3.1. Hence, the cdf of the time interval between the 0th time instant till the first change in vehicle's one-hop neighborhood is given by

$$F_{T_{VN0_r}}(m) = \frac{1}{\sum_{\substack{j \\ \Omega'_j \in \Omega_I}} \mho_j} \sum_{\substack{i \\ \Omega'_i \in \Omega_I}} \frac{\pi_{(i, s_{N_{VN}})} \mho_i F_{T_{VN0_r(\Omega'_i)}}(m)}{\sum_{k=K_i}^{N_{max}} \pi_k} \quad (4.16)$$

where $(i, s_{N_{VN}})$ is the state index of the distance headway of the N_{VN}th VN in the ith lumped state.

---

[2]Since the $(N_{VN} + 1)$-dimensional Markov chain is lumpable into partitions $\Omega_1, \Omega_2, \ldots \Omega_{N_L-1}$, $\Omega_i = \{(s_0, s_1, \ldots, s_{N_{VN}})\}$ contains all super states that are permutations of each other according to Theorem 3.1. Then, it is lumpable into partitions that are subsets of $\Omega_0, \Omega_2, \ldots \Omega_{N_L-1}$.

4.2.2 Time Period Between Successive Changes in Vehicle'S One-Hop Neighborhood

The previous subsection has analysed the time interval from the 0th time instant to the first change in vehicle's one-hop neighborhood change. In order to have a better measure of temporal changes in vehicle's one-hop neighborhood, the time interval between two successive changes in vehicle's one-hop neighborhood is analyzed in this subsection. Let T_{VN} denote the time interval between two consecutive neighborhood changes of a reference vehicle. Notice that the vehicle-neighborhood change rate, i.e. the rate at which nodes enter or leave the vehicle's coverage range, is the reciprocal of T_{VN}. The analysis focuses on one side of the vehicle's coverage range in this subsection, since a similar derivation for the other side can be done.

To derive the distribution of T_{VN}, the first step is to find the distribution of I_{VN} at the time when the first change in vehicle's one-hop neighborhood occurs. In order to do this, first the lumped states in sets Ω_E and Ω_O of the lumped Markov chain are made absorbing, without combining them into one state. The result is an absorbing Markov chain and let $M'_{N_{VN}}$ be its probability transition matrix. Then the probability of absorption in lumped state $\Omega_e \in \Omega_E$ and the probability of absorption in lumped state $\Omega_o \in \Omega_O$ are given respectively by

$$\delta_{E_e} = \frac{1}{\sum_{\substack{j \\ \Omega'_j \in \Omega_I}} \mho_j} \sum_{\substack{i \\ \Omega'_i \in \Omega_I}} \frac{\pi_{(i,s_{N_{VN}})} \mho_i}{\sum_{k=K_i}^{N_{max}} \pi_k} \lim_{m \to \infty} M'^{(m)}_{N_{VN}}(\Omega_i, \Omega_e)$$

and

$$\delta_{O_o} = \frac{1}{\sum_{\substack{j \\ \Omega'_j \in \Omega_I}} \mho_j} \sum_{\substack{i \\ \Omega'_i \in \Omega_I}} \frac{\pi_{(i,s_{N_{VN}})} \mho_i}{\sum_{k=K_i}^{N_{max}} \pi_k} \lim_{m \to \infty} M'^{(m)}_{N_{VN}}(\Omega_i, \Omega_o)$$

where $M'^{(m)}_{N_{VN}}(\Omega_i, \Omega_E)$ denotes the (Ω_i, Ω_E)th entry of the mth power of matrix $M'_{N_{VN}}$. Note that $\sum_{\substack{e \\ \Omega_e \in \Omega_E}} \delta_{E_e}$ and $\sum_{\substack{l \\ \Omega_o \in \Omega_O}} \delta_{O_o}$ are the probabilities that the first change in vehicle's one-hop neighborhood occurs due to a vehicle entering and leaving the coverage range of the reference vehicle, respectively. When calculating the time interval between successive changes in vehicle's one-hop neighborhood, the examined system changes. Let $\mathbb{X}_{VN_E}$ and $\mathbb{X}_{VN_O}$ be the systems of distance headways of the reference vehicle and the nodes on one side of its coverage range when the first vehicle-neighborhood change occurs due to a node entering and a node leaving the vehicle's coverage range, respectively. For example, if system $\mathbb{X}_{VN}$ is absorbed in lumped state $\Omega_i = \{(s_0, s_1, \ldots s_{N_{VN}})\}$, then the initial lumped state for system $\mathbb{X}_{VN_O}$ is $\{(s_0, s_1, \ldots s_{N_{VN}-1})\}$ if $\Omega_i \in \Omega_O$ and the initial lumped state for system $\mathbb{X}_{VN_E}$ is $\{(s_0, s_1, \ldots s_{N_{VN}}, s_{N_{VN}+1})\}$ if $\Omega_i \in \Omega_E$, where $s_{N_{VN}+1} \in [N_R - \sum_{i=0}^{N_{VN}} s_i, N_{max}]$. Let Ω'_e

be the lumped state for system $\mathbb{X}_{VN_E}$ corresponding to lumped state Ω_e for $\mathbb{X}_{VN}$, and let Ω_l' be the lumped state for system $\mathbb{X}_{VN_L}$ corresponding to lumped state Ω_l for $\mathbb{X}_{VN}$. Additionally, let $\delta_{E_e'}$ equal δ_{E_e} weighted by the stationary distribution (3.4) to account for the added distance headway in the system, $\mathbb{X}_{VN_E}$. The cdf of the time interval between two successive changes in vehicle's one-hop neighborhood is approximated by

$$F_{T_{VN}}(m) = \sum_{\substack{\Omega_e \in \Omega_E \\ e}} \delta_{E_e'} F_{T_{VN1e(\Omega_e')}}(m) + \sum_{\substack{\Omega_o \in \Omega_O \\ o}} \delta_{O_o} F_{T_{VN1o(\Omega_o')}}(m). \qquad (4.17)$$

4.3 Temporal Variations in Vehicle'S Two-Hop Neighborhood

In Sects. 4.2.1 and 4.2.2 the time for the first change in vehicle's one-hop neighborhood and the time period between consecutive changes in vehicle's one-hop neighborhood have been investigated. This subsection analyzes the spatiotemporal variations between two-hop vehicles. Consider two non-neighboring reference vehicles that share common VNs, and thus connected via common VNs by two-hop connection. Let N_{H2} be the number of distance headways between two reference vehicles that are two-hop away. The overlapping coverage range between the two reference vehicles is the common distance covered by the transmission range of both vehicles. Define the range-overlap state between a pair of two-hop vehicles to be (1) overlapping, when the distance between the two vehicles is less than $2R$; or (2) non-overlapping, otherwise. Due to vehicle mobility, the total distance between the two reference vehicles can increase, resulting in breakage in the two-hop connection between them. As the distance between the two vehicles increases, diminishing the overlapping range, extra-number of hops need to connect the two reference vehicles. When the coverage range between two-hop vehicles overlap, common VNs within the overlapping range can be used as relay nodes to relay messages from one of the vehicles to the other.

In an overlapping range state, the number of common VNs is an indicator of the strength of the two-hop connection between the two reference vehicles. The larger the number of common VNs, the more candidate relay nodes to connect the two vehicles. On the other hand, in a non-overlapping range state, the number of uncovered nodes, between disjoint coverage ranges of the two reference vehicles, is an indicator of a possible three-hop connection between the two reference vehicles.

The temporal variations in the range-overlap state of two-hop vehicles is governed by the total distance between the two reference vehicles. Label the nodes with IDs: $\{0, 1, \ldots N_{H2}\}$, where the following reference vehicle has ID 0, and the leading reference vehicle has ID N_{H2}. This can be described by the system $\mathbb{X}_{N_{H2}} = \{X_i\}_{i=0}^{N_{H2}-1}$, which is represented by a lumped Markov chain with transition matrix, $M_{N_{2H}}$ derived by applying Theorem 3.1 on the N_{H2}-dimensional

Markov chain. Given initially two-hop connected vehicles, vehicles can enter and leave the overlapping region between them. Additionally, the range-overlap state may change over time. Therefore, in this section, the system of two-hop vehicles is investigated in terms of the change of the numbers of common VNs and uncovered nodes between the two coverage ranges along with the change in the range-overlap state.

The vehicles' coverage ranges remain overlapping until $\sum_{i=0}^{N_{H2}-1} X_i(m) \geq 2R$ at some time step m. Let T_{ov} and T_{nov} be the time periods the coverage ranges of the two reference vehicles remain overlapping and non-overlapping, respectively. The distribution of T_{ov} and T_{nov} can be derived by the same steps used to derive the communication link duration in Sect. 4.1.2.2 by substituting the absorbing set Ω_R by $\Omega_{2R} = \{\Omega_i\}$, s.t., $\Omega_i = \{s_0, s_1, \ldots, s_{N_{H2}-1}\}$ and $\sum_{i=0}^{N_H-1} s_i \geq 2N_R$ and by making the necessary adjustments to Ω_I, $\tilde{N}_L$, and $\tilde{M}_{N_{2H}}$. Consequently, the average time period the coverage ranges of two-reference vehicles remains overlapping is given by

$$E[T_{ov}] = \Phi \left(I - \tilde{M}_{N_{H2}} \right)^{-1} M_1 \qquad (4.18)$$

where Ψ is a row vector of size $\tilde{N}_L$ in which the jth element equals $\phi_j \approx \dfrac{\mathfrak{V}_j \tilde{M}_{N_{H2}}(\Omega_j, \Omega_{\tilde{N}_L-1})}{\sum\limits_{\substack{j \\ \Omega_j \in \Omega_I}} \mathfrak{V}_j \tilde{M}_{N_{H2}}(\Omega_i, \Omega_{\tilde{N}_L-1})}$, I is the identity matrix of size equal to that of $\tilde{N}_L$, and M_1 is a column vector of ones with size $\tilde{N}_L$. The average non-overlapping period $E[T_{nov}]$ can be calculated by (4.18) by substituting the values of $\Phi, I, \tilde{M}_{N_{H2}}$, and M_1 with ones corresponding to an absorbing lumped Markov chain representing the system $\mathbb{X}_{N_{H2}}$, when the lumped states in Ω_I are merged into a single absorbing state.

Since the system of distance headways between the two-hop vehicles, $\mathbb{X}_{N_{H2}}$, constructs a finite irreducible lumped Markov chain, there exists an infinite sequence of range-overlapping and range-non-overlapping time periods [6]. Therefore, the range-overlap state between the reference vehicles fluctuates between overlapping and non-overlapping scenarios. Let $\{\theta(m), m = 0, 1, \ldots\}$ be a stochastic process with state space $\{-1, 1\}$. If $\sum_{i=0}^{N_{H2}} X_i(m) < 2R$, i.e., the reference vehicle's ranges overlap, then $\theta(m) = -1$; otherwise, $\theta(m) = 1$. Denote by $\zeta_1, \eta_1, \zeta_2, \eta_2, \ldots$ the lengths of successive intervals spent in states -1 and 1, respectively, where $\zeta_1, \zeta_2, \ldots$ are i.i.d. and $\eta_1, \eta_2, \ldots$ are i.i.d.. The process $\{\theta(m)\}$ alternates between states -1 and 1, as shown in Fig. 4.5, which is referred to as alternating renewal process [7]. Since the coverage ranges of the reference vehicles are assumed to be initially overlapping, then $\eta(0) = -1$ and $\zeta_k = T_{ov}^k$, and $\eta_k = T_{nov}^k$, i.e., kth range-overlapping period and the kth range-non-overlapping period, respectively. The T_{ov}^k's are assumed to be i.i.d. and the T_{nov}^k's time periods are i.i.d. and they are independent of one another.[3] The kth cycle is composed of ζ_k and θ_k.

---

[3]Index k is dropped from T_{ov}^k and T_{nov}^k to refer to an arbitrary overlapping and non-overlapping period, respectively.

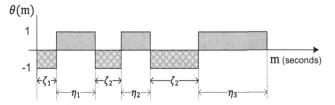

Fig. 4.5 Illustration of the alternating renewal process between overlapping and non-overlapping time periods

4.3.1 Node Interarrival Time During an Overlapping/Non-overlapping Period

During an overlapping/non-overlapping period, vehicles enter and leave the overlapping/uncovered region resulting in a change in the number of common/uncovered nodes between the reference vehicles. Consider two-hop vehicles with their coverage range overlapping. A vehicle entering the overlapping region can be a VN of either of the reference vehicles. Let T_i and T_{Ii} be the first arrival time and the interarrival time of nodes to the overlapping region, respectively. The arrival times considered here are these that cause an increase in the number of common VNs of both reference vehicles. The time for the first node entering the overlapping coverage range is $T_i = \min(T(e_{i_r1}), T(e_{i_l2}))$, where e_{i_r1} is the event that a vehicle enters the following vehicle's coverage range from its right side, and e_{i_l2} is the event that a vehicle enters the leading vehicle's coverage range from its left side as illustrated in Fig. 4.6a. Note that $T(e_{i_r1})$ and $T(e_{i_l2})$ have the same probabilistic behaviors.

The times $T(e_{i_r1})$ and $T(e_{i_l2})$ are assumed to be independent. The times, $T(e_{i_r1})$ and $T(e_{i_l2})$ can then be calculated independently by applying the first passage time analysis on two edge lumped Markov chains, each identifying the hop edge node of its corresponding reference vehicle, as in Sect. 4.2.1. However, The distributions of $T(e_{i_r1})$ and $T(e_{i_l2})$ can be approximated by calculating them from a fully lumped Markov chain with the initial distribution calculated from the state space of the edge lumped Markov chain [12]. Since the distributions of $T(e_{i_r1})$ and $T(e_{i_l2})$ are the same, the following focuses on one of them only, say $T(e_{i_r1})$. Let Θ_E be a set of states of the edge lumped Markov chain for a vehicle with $N_{VN} - 1$ neighbors, such that a lumped state $\Omega_i = \{s_0, s_1, \ldots, s_{N_{VN}}\}$ belongs to Θ_E if $\sum_{i=0}^{N_{VN}-1} s_i < N_R$ and $\sum_{i=0}^{N_{VN}} s_i \geq N_R$. Let $\{\pi_{E,i}\}_{i=1}^{|\Theta_E|}$ be the stationary distribution of the edge lumped Markov chain. Furthermore, divide the lumped states of the fully lumped Markov chain representing system $\mathbb{X}_{VN}$ into two sets, Ω_R and Ω_{R^c}. A lumped state $\Omega_i = \{s_0, s_1, \ldots, s_{N_{VN}}\}$ belongs to Ω_R if $\sum_{i=0}^{N_{VN}} s_i < N_R$ and to Ω_{R^c} otherwise. Let $T(e_{i_r1}, \Omega_k)$ be the first occurrence time of event e_{i_r1} given that system $\mathbb{X}_{VN}$ is initially in lumped state $\Omega_k \in \Omega_R$. Using the recursive formula (4.7),

a

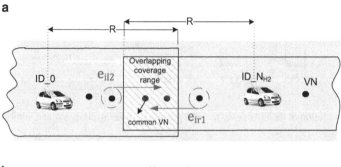

b

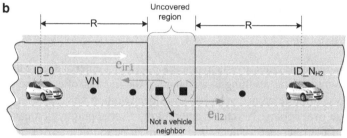

Fig. 4.6 Illustration of the events that cause a vehicle to (**a**) enter the overlapping region and (**b**) leave the uncovered region between two reference vehicles

$F_{T(e_{i_r1},\Omega_j)}(m) = \tilde{M}_{N_{VN}}(\Omega_j, \Omega_{\tilde{N}_L-1}) + \sum_{\Omega_k \in \Omega_R}^{k} \tilde{M}_{N_{VN}}(\Omega_j, \Omega_k) F_{T(e_{i_r1},\Omega_k)}(m-1)$. The cdf of $T(e_{i_r1})$ is approximated by

$$F_{T(e_{i_r1})}(m) \approx \sum_{\substack{j \\ \Omega_j \in \Omega_R}} \omega_j F_{T(e_{i_r1},\Omega_j)}(m), \quad m \geq 1 \tag{4.19}$$

where $\omega_j = \sum_{\substack{i \\ O_E(\Omega_i)=\Omega_j}} \pi_{E,i}$ is the initial probability distribution of states $\Omega_j \in \Omega_R$ and $O_E(\Omega_i)$ is a function that maps a lumped state from edge lumped Markov chain to the corresponding one in the fully Markov chain, note that $\omega_j = 0$ if $\nexists \Omega_i \in \Theta_E$ $s.t. O_E(\Omega_i) = \Omega_j \ \forall \Omega_i \in \Theta_E$ and $\Omega_j \in \Omega_R$.

In order to calculate the probability distribution of node interarrival time to the overlapping region, the probability distribution of the state of the system when a node first enters the vehicle's coverage range needs to be calculated. Consider a vehicle with $N_{VN} - 1$ VNs at time zero. When a node enters the vehicle's coverage range, system $\mathbb{X}_{VN}$ representing the N_{VN} distance headways can only be in an edge lumped state $\Omega_i = \{s_0, s_1, \ldots, s_{N_{VN}}\}$ s.t. the first N_{VN} states construct a lumped state, $\Omega_k = \{s_0, s_1, \ldots, s_{N_{VN}-1}\}$, in a fully lumped Markov chain for system $\mathbb{X}_{VN}$, that satisfies (1) $\Omega_k \in \Omega_R$ and (2) $\tilde{M}_{N_{VN}}(\Omega_k, \Omega_{\tilde{N}_L-1}) > 0$. That is, $\Omega_k \in \Omega_R$ is directly accessible from a lumped state in Ω_{R^c}. As a result, the node interarrival time to the overlapping region from one vehicle's neighborhood can be approximated by

$$F_{T(e_{lir1})}(m) \approx \sum_{\substack{j \\ \Omega_j \in \Omega_R}} \omega_{I_j} F_{T(e_{ir1}, \Omega_j)}(m), \quad m \geq 1 \tag{4.20}$$

where $\omega_{I_j} = \dfrac{\sum_{\substack{i \\ O_E(\Omega_i)=\Omega_j}} \tilde{M}_{N_{VN}-1}(\Omega_i, \Omega_{\tilde{N}_L-1}) \pi_{E,i}}{\sum_{\substack{j \\ \Omega_j \in \Omega_R}} \sum_{\substack{i \\ O_E(\Omega_i)=\Omega_j}} \tilde{M}_{N_{VN}-1}(\Omega_i, \Omega_{\tilde{N}_L-1}) \pi_{E,i}}$ is the probability distribution of the

initial state when a node just entered the vehicle's coverage range, and $\tilde{M}_{N_{VN}-1}$ is the probability transition matrix of the absorbing lumped Markov chain that represents system $\{X_0, X_1, \ldots, X_{N_{VN}-1}\}$. The cdf of the node interarrival time to the overlapping region is given by

$$F_{T_{Ii}}(m) = 1 - (1 - F_{T(e_{lir1})}(m))^2. \tag{4.21}$$

When the two-hop connection between the reference vehicle breaks due to coverage ranges becoming disjoint, vehicles enter and leave the uncovered region. Let us consider the node interdeparture time from the uncovered region that causes the number of uncovered nodes to decrease, denoted by T_{Io}. Nodes can leave the uncovered region and enter either of the two coverage ranges to become VNs. It can be concluded that the time for a node to leave the uncovered region is equal to the minimum of two time intervals $T(e_{i_r 1})$ and $T(e_{i_l 2})$, as illustrated in Fig. 4.6b. Notice that the events that cause the node departure from the uncovered region during a non-overlapping period are the same as those causing the node arrival to the overlapping region during the overlapping period. Therefore, the distribution of T_{Io} can be calculated accordingly.

4.3.2 Steady-State Distributions of the Numbers of Common VNs and Uncovered Nodes

This subsection investigates the limiting behavior of the neighborhood of a two-hop vehicles. Two questions are to be answered: *After a long time, what is the probability that the coverage ranges of the two vehicles is overlapping (non-overlapping)? What is the probability distribution of the number of common VNs (uncovered nodes) in the overlapping (uncovered) region?* By answering the two questions, the limiting probability of the two-hop connection between the two reference vehicles being connected or disconnected can be obtained.

The first question can be answered using the theory of alternating renewal process. The limiting overlapping and non-overlapping probability is given by $P_{ov} = \frac{E[T_{ov}]}{E[T_{ov}]+E[T_{nov}]}$ and $P_{nov} = \frac{E[T_{nov}]}{E[T_{ov}]+E[T_{nov}]}$, respectively [7]. For the second question, the problem can be modelled as a storage buffer with a two-state random environment [8]. The buffer content represents the number of nodes in the overlapping/uncovered region between the two reference vehicles. The two random states of the buffer are the overlapping and the non-overlapping states which fluctuate according to the alternating renewal process as described earlier. Let $N_i(\zeta_k)$ $(N_o(\theta_k))$ be the numbers of nodes entering (leaving) the buffer during the kth overlapping period (non-overlapping period), respectively. Let $N_i(\Delta t)$ $(N_o(\Delta t))$ be the numbers of nodes entering (leaving) the overlapping (uncovered) region during an arbitrary time period, Δt, respectively. The numbers $N_i(\Delta t)$ and $N_o(\Delta t)$ are point processes corresponding to the i.i.d. interrenewal periods T_{Ii} and T_{Io}, and representing the input process (output process) of nodes to (from) the buffer, respectively. The mean and the variance of the input process during an overlapping period and the output process during a non-overlapping period are given by Cox [7]

$$E[N_i(\zeta_k)] = \frac{E[T_{ov}]}{E[T_{Ii}]}, \quad Var[N_i(\zeta_k)] = \frac{c_{T_{Ii}}^2}{E[T_{Ii}]}E[T_{ov}], \tag{4.22}$$

$$E[N_o(\eta_k)] = \frac{E[T_{nov}]}{E[T_{Io}]}, \quad Var[N_o(\eta_k)] = \frac{c_{T_{Io}}^2}{E[T_{Io}]}E[T_{nov}], \tag{4.23}$$

respectively, where $c_{T_{Ii}}$ and $c_{T_{Io}}$ are the coefficients of variation of T_{Ii} and T_{Io}, respectively. Consider the kth cycle. The buffer content at the beginning of the cycle is given by[4] $B_k = [B_{k-1} + N_i(\zeta_{k-1}) - N_o(\theta_{k-1})]^+$. Assuming that the processes $N_i(\zeta_{k-1})$ and $N_o(\theta_{k-1})$ are non-decreasing for all k, the buffer content model can be associated with a G/G/1 queue [8]. In the queueing model, the service time of customer $k - 1$ is $\xi_{k-1} = N_i(\zeta_{k-1})$ and the interarrival time between customers $k - 1$ and k is $A_{k-1} = N_o(\eta_{k-1})$. Then the buffer content at the beginning of the kth cycle is the waiting time of the kth customer. Therefore, the buffer content at an arbitrary time step, m, is equal to the virtual waiting time (or the workload) of this G/G/1 queue [8, 9]. The virtual waiting time depicts the remaining service time of all customers in the system at an arbitrary time step. Let $V(m)$ denote the virtual waiting time (buffer content) at an arbitrary time step m. The relation between the virtual waiting time at the mth time step and the customer waiting time at the beginning of a cycle is given by Kella and Whitt [8]

$$V(m) = \left[B_{n(m)} + \xi_{n(m)} - m + \sum_{k=1}^{n(m)-1} A_k \right]^+ \tag{4.24}$$

where $n(m) = \max\{k \geq 0 : \sum_{i=1}^{k} A_k \leq m\}, m \geq 0$.

To find the limiting probability distribution of the buffer content (i.e., the number of common VNs (uncovered nodes) between two reference vehicles) a diffusion approximation is used. The diffusion approximation is a second order-approximation

---

[4]$y = [x]^+$ is equivalent to $y = \max(0, x)$

that uses the first two moments of the service and interarrival times of the G/G/1 queue [10]. Let $\rho = E[\xi_k]/E[A_k]$ be the intensity factor. A steady-state distribution of the buffer content exists if $\rho < 1$ and it is approximated by a geometric distribution with parameter equal to $\left(1 - \frac{\lambda_g^2}{\lambda_g^2 - 2\chi_g}\right)$. The approximated pmf is given by Kleinrock [10], Kobayashi and Mark [11], and Abboud and Zhuang [12]

$$P_V(n) \approx \left(1 - \frac{\lambda_g^2}{\lambda_g^2 - 2\chi_g}\right)\left(\frac{\lambda_g^2}{\lambda_g^2 - 2\chi_g}\right)^n, \quad n \geq 0 \tag{4.25}$$

where $\chi_g = \rho - 1$ and $\lambda_g^2 = \frac{E[\xi_k^2]}{E[A_k]}$ which can be calculated from (4.22) and (4.23). The limiting probability distribution of the numbers of common VNs and uncovered nodes between the two reference vehicles can be described by the pmf (4.25) with probability P_{ov} and P_{nov}, respectively. Let P_{C0} and P_{U0} denote the limiting probabilities that there are zero common VNs and zero uncovered nodes between the two reference vehicles, respectively. These probabilities are given by $P_{C0} = P_{ov}P_V(0) + P_{nov}$, and $P_{U0} = P_{nov}P_V(0) + P_{ov}$.

4.4   Results and Discussion

This section presents numerical results for the analysis of the spatiotemporal variations in VANET topology in terms of the communication link length, the communication link lifetime, and the vehicle's (one-hop/two-hop) neighborhood changes. Three traffic flow conditions are considered: uncongested, near-capacity, and congested, each corresponding to a set of parameters listed in Table 4.1. The standard deviation of the distance headway is set to $\sigma = \frac{1000}{2D}$ for the mesoscopic distance headway models. The parameters for the microscopic Markov-chain distance headway model are also listed in Table 4.1, where β, p and q follow the VISSIM data fitting results in Sect. 3.1. Without loss of generality, $\alpha = 0$, and $X_{max} = R$. This is sufficient for communication link analysis, as the link breaks if

Table 4.1 System parameters in simulation and analysis of Chap. 4

Traffic flow condition	D (veh/km)	N_H	β	p, q	I_0
Uncongested	9	0	0.4	0.17	{5}
Near-capacity	26	3	0.74	0.23	{1,1,1,1}
Congested	42	5	0.94	0.35	{1,1,1,1,1,1}
R (m)	N_R	N_{max}	α	τ (s)	L_s (m)
160	8	9	0	2	20

any X_i's reach state N_R. The values of N_H and I_0, listed in Table 4.1, are first set to their average values. To verify the spatiotemporal topology analysis, two types of vehicle mobility simulations are conducted:

1. **VISSIM simulations**: A three-lane highway traffic is simulated using the microscopic vehicle traffic simulator VISSIM as described in Sect. 3.1. The choice of simulating a three-lane highway instead of a single-lane highway is to achieve a more realistic vehicle mobility in which a vehicle can overtake other vehicles and accelerate towards its desired speed. The desired speed for all vehicles is normally distributed with mean 100 km per hour and standard deviation of 10 km per hour. Six 30-min simulations are obtained for each of the three vehicle densities

2. **MATLAB simulations**: A set of time series of distance headway data is generated according to the microscopic individual mobility model proposed in Sect. 3.1, using MATLAB. The mobility model describes the time variations of the distance headway between any two consecutive vehicles according to a discrete-time Markov chain model. Each simulation consists of 15,000 iterations. Table 4.1 also lists the parameters of the Markov-chain distance headyway model and the transition probabilities which are tuned according to the results in Sect. 3.1.1.

Figure 4.7 plots the pdfs (4.2)–(4.4) of the hop length for three vehicle densities $D = 9, 26$, and 42 veh/km with average hop length equal to $99.2, 121.8$, and 132.2 m, respectively. The average length of the communication link is larger for a higher vehicle density, due to a larger average number of nodes between a node

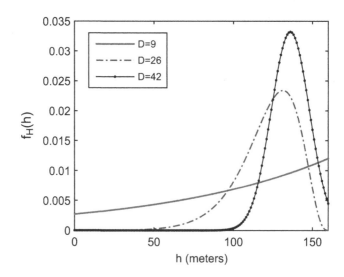

Fig. 4.7 Probability density function of the hop length for three traffic flow conditions with vehicle densities of 9, 26, and 42 veh/km

and its hop edge node. Using (2.1)–(2.3), the probability for an unavailable link between two vehicles (i.e., $P(X_i > R)$) is $0.23, 5.2 \times 10^{-5}$, and 0 for $D = 9, 26$, and 42 veh/km, respectively. That is, the probability of network fragmentations is higher in an uncongested traffic flow condition than that in a congested traffic flow condition.

To verify the link lifetime analysis, the distribution of the communication link lifetime from an initial state, $(T_{R0(I_0)})$, calculated with (4.6) and (4.7) is compared to that calculated from VISSIM and MATLAB simulated vehicular traffic. For the simulation results, the pmf of the lifetime of a link with initial conditions I_0, is calculated by counting the number of occurrences of link breakage at mth time step for $m > 0$ and for all links with initial conditions I_0. The calculation of the link lifetime pmf, $P_{T_{R0(I_0)}}(m)$, from the VISSIM simulated vehicle traffic data includes the lifetime of the following: (1) a link between a reference vehicle and its corresponding hop edge node on the same lane, independently of changing hop edge node during the link's lifetime as long as the initial hop edge node remains in the link; and (2) a new link between a reference vehicle and its new hop edge node on the same lane when its previous link breaks. A link which involves a lane change during its lifetime is excluded from the pmf calculation. The frequency of a link lifetime at value l is upper bounded by $\frac{T_{sim}N_S}{l}$, where T_{sim} is the simulation time and N_S is the total number of vehicles in the simulation. The frequency of l-valued lifetime occurrences in VISSIM data is normalized accordingly in the link lifetime pmf calculation.

Figure 4.8 plots the pmf of the communication link lifetime for the three traffic flow conditions given that the system $\mathbb{X}_{N_H}$ is in state I_0 at the 0th time step. The theoretical results are obtained using (4.6) for the uncongested traffic flow condition and (4.7) for the near-capacity and congested traffic flow conditions. MAPLE software is used to calculate the mth derivative for the generating function in (4.6). For large values of j and/or j' in (4.6), the mth derivative of the product rule proposed in [13] is used. The simulation results are calculated from the generated VISSIM and MATLAB vehicle trajectory data. The simulation results closely agree with the theoretical calculations. However, there exist slight differences between VISSIM simulation and theoretical results. This is mainly due to lane changes, which are not explicitly accounted for in the proposed model. The effect of lane changes is more notable in the low vehicle density simulation results, where zero probability is obtained for some large link lifetime values, as shown in Fig. 4.8a. This is due to the high probability of lane change for large link lifetimes, which is excluded from the calculations. The average link lifetime is found to be 335.5, 88.1, and 65.9 s for the low, intermediate, and high vehicle densities, respectively. Although, intuitively, it is thought that a communication link lasts longer with a higher vehicle density, the results presented here indicate the opposite. The reasons are: (1) the impact of a larger number of vehicles within the link, N_H, with a higher vehicle density and therefore multiple mobility factors on the communication link lifetime; and (2) vehicles tendency to move with their maximum desired speed in an uncongested traffic flow conditions. Since the communication link disconnects when the sum of any of the N_H distance headways is greater than R, the larger the N_H value,

Fig. 4.8 Probability mass function of the communication link lifetime from the 0th time step given that $\mathbb{X}_{N_H}(0) = I_0$ for $D =$ (**a**) 9, (**b**) 26, and (**c**) 42 veh/km

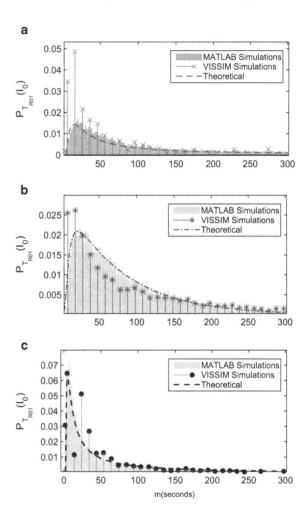

the more frequently a link breakage occurs, for the same traffic flow condition. Although distance headways are large in a low vehicle density scenario with free driving (Table 1.1), this does not necessarily indicate a large probability of changing speeds (i.e., large p and q). On the contrary, vehicles are more likely to be at their maximum desired speeds, resulting in small p and q values [14]. In a congested traffic flow condition, vehicles are more likely to undergo stop-and-go situations, in which drivers speed up whenever they get an opportunity (i.e., large p and q values). This agrees with VISSIM results shown in Fig. 3.4.

From the results shown in Figs. 4.7 and 4.8, the following is concluded: For a high traffic density, there is a higher probability of link availability between two nodes (Fig. 4.7); however, the link lifetime is shorter (Fig. 4.8). This causes the communication link to fluctuate between connection and disconnection more frequently when compared to that in a low vehicle density. This is due to the stop-

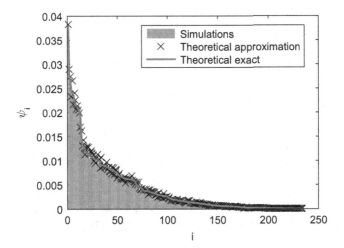

Fig. 4.9 Probability mass function, $\psi_i = P(I_0 \in \Omega_i)$, of system $\mathbb{X}_{N_H}$ being in lumped state $\Omega_i \in \Omega_I$ at the instant when the second connected link-state occurs

and-go scenario in a high vehicle density. On the other hand, for a low traffic density, there is a lower probability of link availability between two nodes (Fig. 4.7); however, if a link exists, the link lasts longer when compared to the case in a high vehicle density (Fig. 4.8). Therefore, when a communication link disconnects in an uncongested traffic flow condition, it has a smaller probability to re-connect than that in a congested traffic flow condition. Recall from Sects. 4.1.1 and 4.1.2.1, that the first link lifetime, depends on the initial conditions N_H and I_0.

In the following, numerical results are presented for the analysis of the proposed analysis of the communication link lifetime given the system $\mathbb{X}_{N_H}$ is in a random initial state I_0, i.e., T_{R0} and the communication link duration from the time two vehicles first become VNs until the time they get disconnected, i.e, T_R. Figure 4.10a plots the pmf of the time for the first link breakage, when averaging over random initial states, for a near capacity vehicle traffic flow condition. The theoretical results for the pmfs of the communication link lifetime is calculated from the cdf in (4.11).

Figure 4.9 compares the distribution of the state of system $\mathbb{X}_{N_H}$, when the second connected link-state occurs, calculated using the exact derivation (4.10) and the proposed approximation (4.12). The values on the x-axis represent arbitrary IDs given to the lumped states $\Omega_i \in \Omega_I$. The results from the proposed approximation shows close agreement with the exact and the simulation results.

Figure 4.10b plots the pmf of the communication link duration from the time the two vehicles gets connected until they get disconnected for a near capacity vehicle traffic condition. The theoretical results for the pmf of the communication link duration, T_R are calculated from the cdf in (4.11). The calculated pmf of T_R in Fig. 4.10b is based on the approximation given in Fig. 4.9. Note that the distribution of $T_{R01}(\Omega_k)$ changes with I_0 belonging to different lumped states Ω_k, as shown in Fig. 4.8b. The distribution of T_{R0} describes the average time before the first link

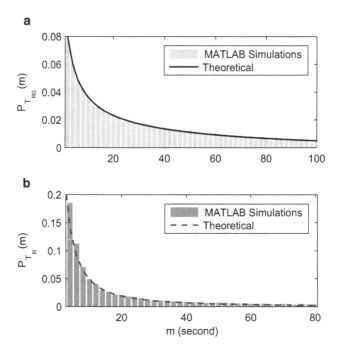

Fig. 4.10 Probability mass functions of (**a**) the time to the first communication link breakage T_{R0}; and (**b**) the communication link duration, T_R, when $D = 26$ veh/km

disconnection for a randomly picked pair of vehicle and its hop edge node in the network. When a link is first established between two nodes, their communication link duration is equal to the time period between two successive link-state changes (i.e., the time period of invariant link-state). Note that the average time for the first change of link-state is larger than the average time period between successive changes of link-state. When the second connected link-state occurs between a vehicle and its hop edge node, the link-state is closer to disconnection than that in the initial time step, on average. That is, the link-state can only be in the accessible lumped states (Ω_{l1} in Fig. 4.2).

In the following, the numerical results of the spatiotemporal variations in vehicle's one-hop neighborhood is presented. Figure 4.11 plots the pmf of the time period from the initial time step till the time step that a first change in vehicle-neighborhood occurs for (a) a given initial state $I_{VN} \in \Omega_k$ and (b) a random initial state. Figure 4.11c plots the pmf of the time period between successive vehicle-neighborhood changes. The theoretical results for the pmfs of $T_{VN0}(\Omega_k)$ and T_{VN0} are calculated from the cdfs in (4.15) and (4.16), respectively. The pmf of the time period between successive changes in vehicle's one-hop neighborhood is calculated from the cdf in (4.17) and is plotted in Fig. 4.11. Figure 4.12 plots the pmfs of the time interval between two successive changes in vehicle's one-hop neighborhood for different vehicle densities. The simulation results closely agree

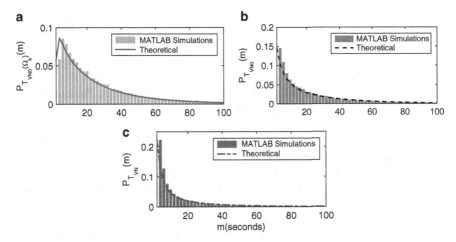

Fig. 4.11 Probability mass functions of (**a**) the time to the first change in vehicle's one-hop neighborhood, $T_{VN0}(\Omega_k)$, for $I_{VN} = \{1, 1, 1, 1, 5\} \in \Omega_k$, (**b**) the time to the first change in vehicle's one-hop neighborhood, T_{VN0}; and (**c**) the time period between two successive vehicle's one-hop neighborhood changes, T_{VN}, when $D = 26$ veh/km

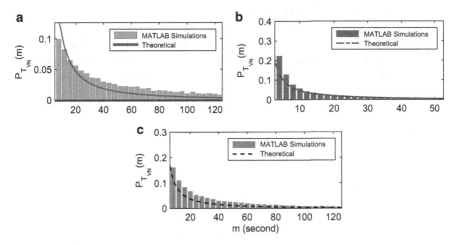

Fig. 4.12 Probability mass function of the time period between successive changes vehicle's one-hop neighborhood with vehicle density (**a**) $D = 9$, (**b**) $D = 26$, and (**c**) $D = 42$ veh/km

with the theoretical calculations. It is observed that, when the first change in vehicle's one-hop neighborhood occurs from 0th time, the second change in vehicle's one-hop neighborhood has a higher probability of occurring in a shorter time period. This reflects the effect of a wireless link between a reference vehicle and its neighbors fluctuating between connecting and disconnecting states in a short period of time. The impact of this fluctuation can lead to frequent resource allocations and packet collisions, that drains the precious VANET radio resources.

For the spatiotemporal variations in a two-hop vehicle neighborhood analysis, firstly, the two-hop reference vehicles at the initial time (0th time step) are selected with a simple selection algorithm. The distance headways of vehicles on the highway follow a truncated exponential, gamma, and Gaussian distributions for the uncongested, near-capacity, and congested traffic flow conditions, respectively. The vehicles' speeds are i.i.d. and are normally distributed with mean 100 km per hour and standard deviation of 10 km per hour [14]. The potential reference vehicles are selected with the minimum average relative speed to its one-hop neighbors, such that each node is a VN of a reference vehicle and no two reference vehicles are one-hop neighbors. The resulting set of reference vehicles are two-hop away and their coverage ranges overlap, which satisfies the assumption for the initial two-hop reference vehicles. Figure 4.13 plots the probability distribution of the number of distance headways between two-hop vehicles, N_{H2}, for the resulting set of reference vehicles from simulating the selection algorithm. Figure 4.14 plots the pmf of the interarrival time period of nodes into the overlapping region. Figure 4.15 plots the average range-overlapping and the average range-non-overlapping time periods for different numbers of nodes between two reference vehicles that are initially two-hop away, N_{H2}. The average values are calculated using (4.13) and

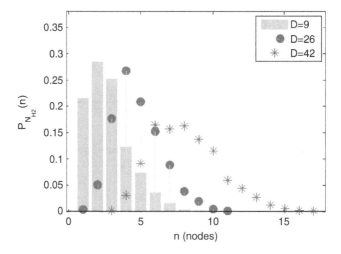

Fig. 4.13 Probability mass function of the number of nodes between two-hop reference vehicles, N_{H2} calculated from simulating a simple selection algorithm

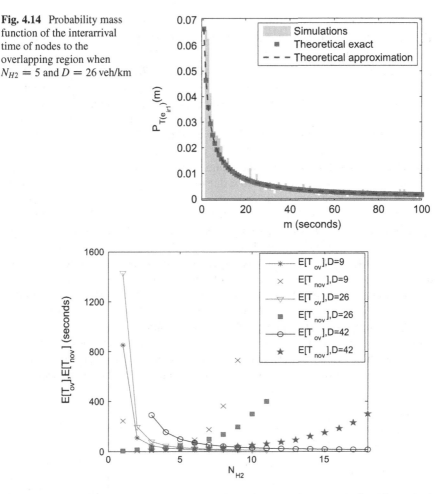

Fig. 4.14 Probability mass function of the interarrival time of nodes to the overlapping region when $N_{H2} = 5$ and $D = 26$ veh/km

Fig. 4.15 Average range-overlapping and range-non-overlapping time periods for different N_{H2} values with vehicle density $D = 9, 26$, and 42 veh/km. The values of N_{H2} are those in Fig. 4.13

the values of N_{H2} are from the selection results in Fig. 4.13. For a fixed N_{H2}, the average range-overlapping period is larger for a larger density, whereas the average range-non-overlapping period is smaller for a larger density. The reason is that, in a congested traffic flow conditions, the distance headways are small when compared to those in an uncongested traffic flow condition. Therefore, for the same N_{H2}, the cumulative distances are smaller for a high density. It should be noted that the large values of average range-overlapping time periods for $N_{H2} = 1$ are due to the connected network assumption. Figure 4.15 shows that, as N_{H2} increases, the average range-overlapping period reduces and the average range-non-overlapping period increases for the same traffic flow condition.

To investigate the limiting behavior of the number of common VNs in the overlapping region (the nodes in the uncovered region), the two parameters χ_g and

λ_g^2 are first calculated for the three vehicle densities. Notice that the distributions of T_{Ii}, T_{ov} and T_{nov} are all conditional on the initial state of system $\mathbb{X}_{H2}$ in terms of N_{H2} and N_{VN}. Therefore, in the calculation of χ_g and λ_g^2, the law of total expectation is used to calculate $E[T_{ov}] = \sum_n P_{N_{H2}}(n)E[T_{ov}(n)]$ and $E[T_{Ii}^2] = \sum_n P_{N_{H2}}(n)E[T_{Ii}^2(n)]$, where $T_{ov}(n)$ is the range-overlapping time period for two reference vehicles separated by $N_{H2} = n$ nodes and $T_{Ii}(n)$ is the node interarrival time for a vehicle's one-hop neighborhood with $N_{VN} = n$ nodes, respectively. The calculations are done for near-capacity and congested traffic flow conditions only. The reason is that the diffusion approximation assumes that the point processes $N_i(\zeta_k)$ and $N_o(\theta_k)$ are normally distributed according to the central limit theorem. This assumption is not satisfied for an uncongested traffic flow, due to a relatively small number of vehicles between the two reference vehicles as shown in Fig. 4.13. The intensity factor is found to be $\rho = 1.0143$ and 1.3172 for $D = 26$ and 42 veh/km, respectively. As a result, the steady-state distribution does not exist. However, consider only $N_{H2} \geq E[N_{H2}]$ for both cases, it can be found that $\rho = 0.33$, and 0.64 for $D = 26$, and 42 veh/km, respectively. Figure 4.16 plots the steady-state probability distributions for the non-zero number of vehicles in the overlapping/uncovered region when $N_{H2} \geq E[N_{H2}]$ for near-capacity and congested traffic flow conditions. The theoretical results are normalized to the value $1 - P_V(0)$, since the probability distributions in Fig. 4.16 represent the non-zero number of common VNs with probability P_{ov} and the non-zero number of uncovered nodes with probability P_{nov}. The simulation results closely agree with the theoretical calculations. However, there exist slight differences between simulation and theoretical results especially

Fig. 4.16 Steady-state pmf of the buffer content, i.e., the number of non-zero nodes in the overlapping/non-overlapping period, when (a) $D = 26$ and (b) $D = 42$ veh/km

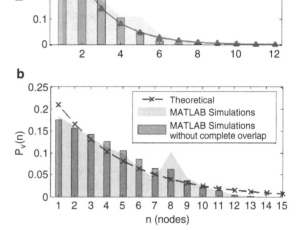

Table 4.2 Limiting
probabilities of zero common
VNs/uncovered nodes

	D (veh/km)	Simulation	Theoretical
P_{C0}	26	0.86	0.78
P_{C0}	42	0.69	0.67
P_{U0}	26	0.57	0.47
P_{U0}	42	0.52	0.56

at the values of $n = 5$ and $n = 8$, for $D = 26$ and 42 veh/km, respectively. This is mainly due to complete overlapping of the coverage ranges of the reference vehicles (i.e., the reference vehicles becoming one-hop neighbors). That is, the reference vehicles become one-hop neighbors and all the nodes between them become common VNs. As a result, no additional nodes can enter the overlapping region when its completely overlapping. This is not accounted for in the presented two-hop vehicle neighborhood analysis. In this case, the spatiotemporal analysis of one-hop vehicle neighborhood described in Sect. 4.2 applies. Figure 4.16 shows that the simulation results excluding the complete range-overlapping data are in closer agreement with the theoretical results in comparison with simulation results that include the complete range-overlapping data. Additionally, the numerical and simulation results for the limiting probabilities of having zero common VNs and zero uncovered nodes are given in Table 4.2.

4.5   Summary

This chapter presents a stochastic analysis of spatiotemporal variations in highway VANET topology with focus on a single lane. The communication link duration and the time periods between consecutive changes in vehicle's one-hop neighborhood are proposed as measures of topological changes in VANETs. Stochastic mobility models that describe the time variations of an individual distance headway and a group of distance headways are adopted in the analysis. The system of distance headways that govern the changes in communication links is modeled by a discrete time lumped Markov chain. The first passage time analysis is employed to derive the distributions of the proposed measures of topological changes in VANETs. Additionally, the limiting probability distributions of the numbers of common vehicle neighbors and uncovered nodes between two-hop vehicles are approximated using queuing theory and diffusion approximation. The probability distributions derived for the measures of network topological changes can be utilized in the development of efficient mobility-aware protocols for VANETs.

References

1. T. Hou and V. Li, "Transmission range control in multi-hop packet radio networks," *IEEE Trans. Communications*, vol. 34, no. 1, pp. 38–44, 1986.
2. Y. Cheng and T. Robertazzi, "Critical connectivity phenomena in multihop radio models," *IEEE Trans. Communications*, vol. 37, no. 7, pp. 770–777, 1989.
3. K. Abboud and W. Zhuang, "Stochastic analysis of single-hop communication link in vehicular ad hoc networks," *IEEE Trans. Intelligent Transportation Systems*, vol. 15, no. 5, pp. 2297–2307, 2014.
4. J. Fill, "The passage time distribution for a birth-and-death chain: Strong stationary duality gives a first stochastic proof," *J. Theoretical Probability*, vol. 22, no. 3, pp. 543–557, 2009.
5. Maple, *Maple (Version 13). Waterloo, Ontario, Canada: Waterloo Maple Software*, 2009.
6. G. Rubino and B. Sericola, "Sojourn times in finite Markov processes," *J. Applied Probability*, vol. 26, no. 4, pp. 744–756, 1989.
7. D. Cox, *Renewal theory*.   Methuen London, 1962.
8. O. Kella and W. Whitt, "A storage model with a two-state random environment," *Operations Research*, vol. 40, no. 3-Supplement-2, pp. S257–S262, 1992.
9. J. Cohen, "Superimposed renewal processes and storage with gradual input," *Stochastic Processes and their Applications*, vol. 2, no. 1, pp. 31–57, 1974.
10. L. Kleinrock, *Queueing Systems, Volume II: Computer Applications*. John Wiely & Sons, 1976.
11. H. Kobayashi and B. L. Mark, *System Modeling and Analysis: Foundations of System Performance Evaluation*.   Prentice Hall, 2009.
12. K. Abboud and W. Zhuang, "Stochastic modeling of single-hop cluster stability in vehicular ad hoc networks," *IEEE Trans. Vehicular Technology*, 2015 (to appear).
13. D. Mazkewitsch, "The n-th derivative of a product," *The American Mathematical Monthly*, vol. 70, no. 7, pp. 739–742, 1963.
14. A. May, *Traffic Flow Fundamentals*. Prentice Hall, 1990.

Chapter 5
Conclusions and Future Work

5.1 Conclusions

VANETs are promising additions to our future intelligent transportation systems, that have captured world-wide attention from auto companies, academics, and government agencies. Realizing V2I and V2V communications will enable many safety and infotainment applications that can revolutionize the transport infrastructure.

VANETs are prone to high speeds, variable relative mobility, traffic jams, and high traffic density. The movements of vehicles with high and variable speeds cause frequent changes in the network topology, which can affect network protocol performance. In order to design network protocols that are robust to network topology changes, the impact of vehicle mobility on VANET topology should be analyzed. This brief presents stochastic analysis of the impact of vehicle mobility on spatiotemporal variations in VANET topology.

A microscopic vehicle mobility model is proposed to describe the time variations of a distance headway. It models the distance headway as a discrete-time Markov chain that preserves the realistic dependency of distance headway changes at consecutive time steps. This dependency is consistent with highway traffic patterns from empirical NGSIM and simulated VISSIM data sets. Furthermore, the proposed microscopic mobility model is mapped into a lumped mobility model that describes the mobility of a group of vehicles.

Due to vehicle mobility, communication links between network nodes switch between connection and disconnection. Changes in communication link-state, caused by vehicle mobility, are major contributors to network topology variations. To capture the changes in VANET topology, firstly, the communication link length and its lifetime are analyzed. The proposed microscopic mobility model is used to analyze the communication link lifetime between a reference node and its hop edge node. Results indicate that on average, congested traffic flow conditions can cause communication links to break faster than in free traffic flow conditions. Secondly,

© The Author(s) 2015
K. Abboud, W. Zhuang, *Mobility Modeling for Vehicular Communication Networks*,
SpringerBriefs in Electrical and Computer Engineering,
DOI 10.1007/978-3-319-25507-1_5

the change in vehicle's one-hop neighbors is analyzed in terms of the time to the first change in vehicle's one-hop neighborhood, and the time period between consecutive changes in vehicle's one-hop neighborhood. In addition, the system of two-hop vehicles is investigated in terms of the change in the number of common VNs between the two vehicles along with the change in their coverage range-overlap state. The limiting behavior of this system is approximately characterized by simple mathematical expressions.

The results of this brief will help to develop guidelines for mobility-aware protocol design in VANETs.

Firstly, The proposed mobility model enables traceable mathematical analysis in VANETs. Furthermore, the proposed lumped group mobility model facilitates the VANET analysis for large vehicle densities (e.g., analyzing communication link lifetime in the case of high number of one-hop neighbors). Secondly, the spatiotemporal analysis is a candidate tool for evaluating the performance of proposed network protocols for VANETs in the presence of mobility, for example: (1) calculating the packet collision probability for a certain MAC design, when vehicles concurrently broadcasting a packet become two-hop neighbors due to mobility (merging collision probability in [1]), (2) characterizing the multi-hop route duration, route availability, and route caching probability for a certain routing protocol design [2], or (3) evaluating single-hop node cluster stability for a proposed clustering algorithm [3].

Despite the simplicity of the system model considered in this brief, the results can be utilized for more general systems. The brief focuses on vehicle traffic flow in one direction only. However, this is applicable to many VANET application scenarios, e.g., safety applications. When a sudden break or an accident occurs, safety messages are generated by the source vehicle. The safety message is disseminated upstream to vehicles traveling in the same direction as the source vehicle [4, 5]. In this case, the safety messages are only relevant to vehicles traveling in the same direction as they are in the danger zone. Although this research work focuses on a single lane only, the lane is considered to be from a multi-lane highway. Therefore, the results of this brief account for the majority of highway traffic scenarios in reality, rather than the limited case of a single lane highway.

This brief has mainly addressed the effect of vehicle mobility, in terms of changing intervehicle distance headways over time, on communication link lifetime and on vehicle's one-hop/two-hop neighborhood. The resulting analyses of this research can still be insightful in the early stages of VANET implementation, i.e., when the penetration rate of the V2V technology is low. This is because the effect of mobility is independent of the penetration rate of the V2V technology. For example, the communication link between two equipped vehicles separated by a number of non-equipped vehicles is the same as that calculated in Chap. 3.

5.2 Future Research Direction

The communication link analysis (and, therefore, the spatiotemporal topology variation analysis), presented in this brief, depends only on the link distance. In reality, the communication link between two nodes depends not only on the distance between the two nodes, but also on the communication channel condition. Although the distance between two nodes may be less than the communication range, poor channel conditions can result in inability of the two nodes to communicate. Both vehicle mobility and vehicle density affect the communication channel conditions. Additionally, as the vehicle density increases to a traffic jam situation, the network data load increases. In this case, the communication between two nodes (and, therefore, the link lifetime) is controlled by the network data traffic congestion rather than by vehicle mobility. Extending the communication link analysis to account for the communication channel condition and the network data load needs further investigation.

To date, VANET analysis and protocol design are mainly based on vehicle mobility models and/or vehicle traffic patterns on urban roads and highways, irrespective of the communication effect on the vehicle traffic. How VANETs implementation affect the vehicle traffic patterns and how to account for the changes in vehicle traffic pattern in the protocol design is an important research topic.

References

1. H. Omar, W. Zhuang, and L. Li, "VeMAC: A TDMA-based MAC protocol for reliable broadcast in VANETs," *IEEE Trans. Mobile Computing*, vol. 12, no. 9, pp. 1724–1736, 2013.
2. K. Abboud and W. Zhuang, "Impact of microscopic vehicle mobility on cluster-based routing overhead in VANETs," *IEEE Trans. Vehicular Technology, connected vehicle series*, 2015 (to appear).
3. K. Abboud and W. Zhuang, "Stochastic modeling of single-hop cluster stability in vehicular ad hoc networks," *IEEE Trans. Vehicular Technology*, 2015 (to appear).
4. K. Abboud and W. Zhuang, "Modeling and analysis for emergency messaging delay in vehicular ad hoc networks," in *Proc. IEEE Globecom*, 2009, pp. 1–6.
5. "Vehicle safety communications project: task 3 final report: identify intelligent vehicle safety applications enabled by DSRC," CAMP Vehicle Safety Communications Consortium, National Highway Traffic Safety Administration (NHSTA), U.S. Department of Transportation, Tech. Rep. DOT HS 809 859, 2005.

Printed in the United States
By Bookmasters